Common Florida Angiosperm Families

Part II

Wendy B. Zomlefer

Front cover: *Nymphaea odorata*, x 1/2

Back cover (clockwise from upper left): Sapotaceae (Fig. 6, pg. 15); Bromeliaceae (Fig. 35, pg. 87); Rhizophoraceae (Fig. 29, pg. 72); Cactaceae (Fig. 9, pg. 22).

ISBN 0-932353-02-9

Published by Biological Illustrations, Inc.
P.O. Box 15292, Gainesville, FL 32604

Printed by Storter Printing Co.
Gainesville, FL U.S.A.

PREFACE

Common Florida Angiosperm Families, Parts I and *II* represent a greatly expanded and modified Master's thesis combining botanical illustration with general summaries of family characteristics. Seventy families (79 original plates) are presented for botanists, students, and other plant enthusiasts. It is hoped that this manuscript will serve as a lab manual in local introductory taxonomy courses and promote appreciation of our Florida flora.

The author assumes all editorial responsibility and would greatly appreciate any corrections, comments, and helpful criticisms.

Wendy B. Zomlefer

Florida State Museum
Natural Sciences Department
University of Florida
Gainesville, FL 32611

June 30, 1985

ACKNOWLEDGEMENTS

I express my deepest appreciation and gratitude to Dr. Walter S. Judd for his guidance and helpful criticisms throughout this project. Kent D. Perkins deserves a very special thank-you for his help in all aspects of the study, including his information on word processing, references, and plant locations, as well as for proofreading the manuscript and accompanying me on countless collecting trips. I am grateful to Dr. E. Eugene Spears, Jr., for helping me to collect appropriate plant material on numerous occasions and to J. Dan Skean, Jr., and Drs. Gerald L. Benny and John W. Thieret for their criticisms of the text.

I thank Howard Adams, Debbie L. White, Elizabeth Conner, James Cobb, Donna Ford, and Drs. Dana G. Griffin, III, Stuart Skeate, and Stephen Sundlof for their help with locating certain plants; Dr. Elizabeth L. Wing for approving my two-month leave of absence from my job at the Florida State Museum (to work on the illustrations for *Part II*); Drs. Michael R. Zomlefer and Norris H. Williams for their technical assistance concerning personal computers; Dr. John Popenoe for allowing me to collect specimens at Fairchild Tropical Gardens; Bobbi Angell and Drs. Daniel F. Austin, David H. Wagner, Bijan Dehgan, James R. Massey, and Andrew M. Greller for their comments and suggestions on *Part I;* Paloma Ibarra, John Knaub, and Ashley Wood for their advice on photography and layout; and Susan W. Williams for her cheerful encouragement.

Finally, I thank my parents, Jack and Dorothy Zomlefer, who have always encouraged and supported my pursuits in both biology and art.

BIOGRAPHICAL SKETCH

Wendy B. Zomlefer has a Bachelor of Science degree from the University of Vermont and a Master of Science degree from the University of Florida. She has twelve years experience as a botanical illustrator, including two years as the staff illustrator for the Marie Selby Botanical Gardens in Sarasota, Florida. For the past four years, she has been the biological illustrator for the Natural Sciences Department of the Florida State Museum at the University of Florida. She is a member of the American Society of Plant Taxonomists, the Florida Native Plant Society, the Guild of Natural Science Illustrators, the Graphic and Scientific Illustrators Association, and the Committee of Small Magazine Editors and Publishers.

Her illustrations have been published extensively in books and scientific journals. The *Common Florida Angiosperm Families* project includes some of her finest pen and ink work.

TABLE OF CONTENTS

INTRODUCTION

Common Florida Angiosperm Families, Part II continues the series initiated in 1983 with *Part I*. In addition to those sources cited in *Part I*, the following references have been useful in compiling the family lists and diagnoses: Hickey and King (1981), Baumgardt (1982), Dahlgren and Clifford (1982), Dahlgren et al. (1985), and Clewell (1985). *How to Draw Plants* by West (1983) is a good source for information on botanical illustration.

Family List for Parts I and II

Seventy families have been chosen on the basis of size and floristic importance within Florida, phylogenetic interest, and economic importance. These families, distributed throughout Thorne's (1983) superorders, are usually covered in introductory taxonomy courses. Thirty-four families with 42 figures were included in *Part I*; the starred (*) 36 families are included in *Part II*.

Class Angiospermae

Subclass Dicotyledonae (Annonidae)

Superorder Annoniflorae
Order Annonales
   (1) **Magnoliaceae**
   (2) **Annonaceae**
   (3) **Lauraceae**
Order Berberidales
  * (4) **Ranunculaceae**
Order Nelumbonales
  * (5) **Nelumbonaceae**

Superorder Nymphaeiflorae
Order Nymphaeales
  * (6) **Cabombaceae**
  * (7) **Nymphaeaceae**

Superorder Theiflorae
Order Theales
   (8) **Aquifoliaceae**
  * (9) **Sarraceniaceae**
   (10) **Clusiaceae** (Guttiferae; including Hypericaceae)
Order Ericales
  (11) **Ericaceae** (including Pyrolaceae and Monotropaceae)
Order Ebenales
  *(12) **Sapotaceae**
Order Polygonales
  (13) **Polygonaceae**

Superorder Chenopodiiflorae
Order Chenopodiales

  *(14) **Caryophyllaceae**
  *(15) **Chenopodiaceae**
  (16) **Amaranthaceae**
  *(17) **Cactaceae**

Superorder Geraniiflorae
Order Geraniales
  *(18) **Oxalidaceae**
  *(19) **Polygalaceae**

Superorder Violiflorae
Order Violales
  (20) **Violaceae**
  *(21) **Salicaceae**
  *(22) **Cucurbitaceae**
Order Capparales
  (23) **Brassicaceae** (Cruciferae)

Superorder Malviflorae
Order Malvales
  (24) **Malvaceae**
Order Urticales
  *(24) **Ulmaceae**
  (26) **Urticaceae** (*sensu stricto* in *Part I*)
  *(27) **Moraceae** (*sensu stricto* in *Part II*)
Order Euphorbiales
  (28) **Euphorbiaceae**

Superorder Rutiflorae
Order Rutales
  (29) **Rutaceae**
  *(30) **Anacardiaceae**

*(31) **Juglandaceae**
*(32) **Aceraceae**
(33) **Fabaceae** (Leguminosae)

Superorder Hamamelidiflorae
Order Fagales
(34) **Fagaceae**
(35) **Betulaceae**

Superorder Rosiflorae
Order Rosales
(36) **Rosaceae**

Superorder Myrtiflorae
Order Myrtales
*(37) **Melastomataceae**
*(38) **Onagraceae**
*(39) **Myrtaceae**

Superorder Gentianiflorae
Order Oleales
*(40) **Oleaceae**
Order Gentianales
(41) **Rubiaceae**
(42) **Apocynaceae** (including Asclepiadaceae)
Order Bignoniales
*(43) **Bignoniaceae**
(44) **Scrophulariaceae**
*(45) **Acanthaceae**
Order Lamiales
*(46) **Verbenaceae**
(47) **Lamiaceae** (Labiatae)

Superorder Solaniflorae
Order Solanales
*(48) **Boraginaceae**
(49) **Solanaceae**
*(50) **Convolvulaceae**
Order Campanulales
*(51) **Campanulaceae**

Superorder Corniflorae
Order Cornales
*(52) **Rhizophoraceae**
*(53) **Vitaceae**

*(54) **Cornaceae**
Order Araliales
(55) **Araliaceae** (including Apiaceae or Umbelliferae)

Superorder Asteriflorae
Order Asterales
(56) **Asteraceae** (Compositae)

Subclass Monocotyledoneae (Liliidae)

Superorder Liliiflorae
Order Liliales
(57) **Liliaceae** (including Amaryllidaceae and Agavaceae)
*(58) **Iridaceae**
(59) **Orchidaceae**

Superorder Alismatiflorae
Order Alismatales
(60) **Alismataceae**
*(61) **Hydrocharitaceae**

Superorder Ariflorae
Order Arales
(62) **Araceae**
*(63) **Lemnaceae**

Superorder Areciflorae
Order Arecales
(64) **Arecaceae** (Palmae)

Superorder Commeliniflorae
Order Commelinales
*(65) **Bromeliaceae**
*(66) **Xyridaceae**
(67) **Juncaceae**
(68) **Cyperaceae**
(69) **Commelinaceae**
(70) **Poaceae** (Gramineae)

Possible further additions to this series include the following families: Theaceae, Rhamnaceae, Portulacaceae, Cistaceae, Sapindaceae, Combretaceae, Gentianaceae, Lentibulariaceae, Lythraceae, and Eriocaulaceae. An index and completely illustrated glossary will also be compiled.

<u>References</u> <u>Cited</u>

Baumgardt, J.P. 1982. How to identify flowering plant families. A practical guide for horticulturalists and plant lovers. 269 pp. Timber Press, Beaverton, OR.

Clewell, A.F. 1985. Guide to the vascular plants of the Florida panhandle. 605 pp. Univ. Presses of Florida, Gainesville, FL.

Dahlgren, R.M.T. and H.T. Clifford. 1982. The monocotyledons: a comparative study. 378 pp. Academic Press, Inc., London, England.

_____, _____, and P.F. Yeo. 1985. The families of monocotyledons. 520 pp. Springer-Verlag, Berlin, Germany.

Hickey, M. and C.J. King. 1981. 100 families of flowering plants. 567 pp. Cambridge Univ. Press, Cambridge, England.

Thorne, R.F. 1983. Proposed new realignments in the angiosperms. Nord. J. Bot. 3: 85-117

West, K. 1983. How to draw plants. 152 pp. Watson-Guptill Publications, New York, NY.

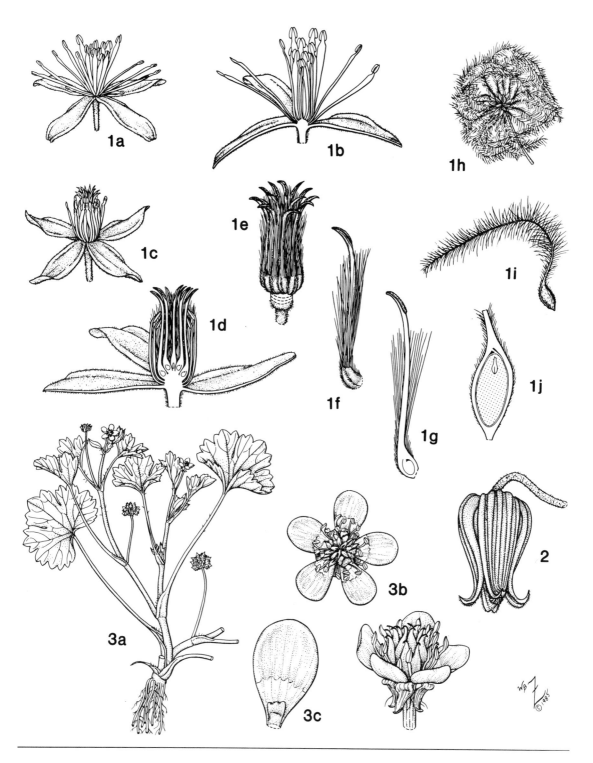

Figure 1. Ranunculaceae. 1, *Clematis catesbyana:* **a,** staminate flower, x 2; **b,** longitudinal section of staminate flower, x 3; **c,** carpellate flower, x 2; **d,** longitudinal section of carpellate flower, x 4; **e,** gynoecium, x 5; **f,** carpel, x 8; **g,** longitudinal section of carpel, x 9; **h,** fruit (aggregate of achenes), x 1; **i,** achene, x 2 1/2; **j,** longitudinal section of achene, x 8. **2,** *Clematis reticulata:* flower, x 1 1/2. **3,** *Ranunculus muricatus:* **a,** habit, x 1/3; **b,** two views of flower, x 3 1/2; **c,** adaxial view of petal with nectary at base, x 6.

RANUNCULACEAE (BUTTERCUP OR CROWFOOT FAMILY)

Annual or perennial herbs, sometimes shrubs or vines, terrestrial or sometimes aquatic, often with rhizomes or tuberous roots. <u>Leaves</u> simple or variously compound or dissected, usually alternate, exstipulate, often with sheathing petioles. <u>Inflorescence</u> determinate, basically cymose and often appearing racemose or paniculate, or flower sometimes solitary, terminal. <u>Flowers</u> actinomorphic or sometimes zygomorphic, usually perfect, hypogynous, often showy, with short and globose to elongate receptacle. <u>Calyx</u> of typically 5 sepals, distinct, caducous, often showy and petaloid, variously colored, imbricate. <u>Corolla</u> of typically 5 petals or often absent, distinct, with modified nectariferous bases or reduced to small nectariferous sacs or scales, imbricate. <u>Androecium</u> of usually numerous stamens spirally arranged on the receptacle, with the outer stamens sometimes reduced to staminodes; filaments distinct; anthers basifixed, dehiscing longitudinally, extrorse. <u>Gynoecium</u> apocarpous and of usually several to numerous carpels spirally arranged on the receptacle; ovary superior, 1-loculed; ovules 1 to numerous per carpel, anatropous, placentation parietal along the ventral suture or nearly basal; style 1; stigma 1, often bilobed. <u>Fruit</u> typically an aggregate of follicles, achenes, or berries; endosperm copious, oily; embryo minute, straight.

Family characterization in Florida: herbaceous plants commonly with rhizomes or tubers; compound or dissected leaves with sheathing petiole bases; perianth often of showy petaloid sepals and reduced or modified petals; numerous distinct stamens and carpels spirally arranged on the receptacle; aggregate fruit of follicles, achenes, or berries; and seeds with minute embryo and copious endosperm. Sap characteristically with various alkaloids, glycosides, and/or saponins.

Genera/species: 46/1,900

Distribution: Primarily in temperate to boreal regions of the Northern Hemisphere; especially diverse in eastern Asia and eastern North America.

Major genera: *Ranunculus* (400 spp.), *Aconitum* (300 spp.), *Clematis* (250 spp.), *Delphinium* (200 spp.), *Anemone* (150 spp.), and *Thalictrum* (150 spp.).

Florida representatives: 11-12 genera/27-28 spp.; largest genera: *Ranunculus* (9 spp.), *Clematis* (7 spp.), *Thalictrum* (3 spp.), and *Delphinium* (2 spp.); see Keener (1975a,b; 1976a,b,c; 1977).

Economic plants and products: Several poisonous and medicinal plants (due to various alkaloids, glycosides, and/or saponins), including: *Aconitum* (wolfbane, monkshood -- with aconite), *Actaea* (baneberry -- with berberine), *Delphinium* (larkspur -- with delphinine), *Hydrastis* (golden-seal -- with berberine), and *Ranunculus* (buttercup -- with ranunculin). Ornamental plants (species of 29 genera), including: *Aquilegia* (columbine), *Anemone* (windflower), *Clematis* (virgin's-bower), *Delphinium* (larkspur), *Helleborus* (hellebore), *Ranunculus* (buttercup, crowfoot), *Thalictrum* (meadow-rue), and *Trollius* (globe-flower).

The Ranunculaceae have generally been regarded as primitive with such features as the numerous carpels and stamens spirally arranged on an enlarged receptacle (see Magnoliaceae in *Part I* and Leppik, 1964). The family is often divided into two or three subfamilies (and several tribes) based on characters of the ovules (number per ovary) and fruit (aggregates of follicles, berries, or achenes). Generic limits (often based on perianth features) tend to be indistinct in this natural but diverse family.

The differentiation of the perianth varies considerably within the family. The flowers of *Ranunculus* have a greenish calyx and showy corolla, as is typical of most flowers. Many genera, such as *Clematis* and *Anemone*, are characterized by apetalous flowers with petaloid sepals. The perianth forms spurs in the more specialized groups, such as *Aquilegia* (spurred petals) and *Delphinium* (spurred sepals and petals).

The conspicuous flowers of this family are usually pollinated by insects. Nectar is often secreted by the petals, such as in *Ranunculus* (nectaries at petal bases) or *Aquilegia* (nectaries within spurs). The flowers of several taxa (e.g., *Clematis* and *Anemone* spp.) that do not produce nectar are visited by insects that gather the pollen. Details of the pollination mechanisms vary greatly depending upon the particular floral morphology. The more open flowers (e.g., those of *Ranunculus*) are visited by various insects, while more elaborate flowers with spurs (e.g., those of *Aquilegia*) are pollinated by insects with long proboscises (or by hummingbirds). The flowers are generally protandrous, although self-pollination is possible when the anthers of the inner whorl dehisce immediately before or during the maturation of the stigmas.

References Cited

Keener, C.S. 1975a. Studies in the Ranunculaceae of the southeastern United States: I: *Anemone* L. Castanea 40: 36-44.
_____. 1975b. Ibid. III: *Clematis* L. Sida 6: 33-47.
_____. 1976a. Ibid. II: *Thalictrum* L. Rhodora 78: 815-472.
_____. 1976b. Ibid. IV: Genera with zygomorphic flowers. Castanea 41: 12-20.
_____. 1976c. Ibid. V: *Ranunculus* L. Sida 6: 266-283.
_____. 1977. Ibid. VI: Miscellaneous genera. Sida 7: 1-12.
Leppik, E.E. 1964. Floral evolution in the Ranunculaceae. Iowa State Coll. J. Sci. 39: 1-101.

NYMPHAEACEAE (WATER-LILY FAMILY)

Perennial aquatic herbs, scapose, with milky sap; rhizome large, with massive shoot apices. <u>Leaves</u> simple, usually entire, alternate, with palmate to pinnate venation, long-petiolate with usually floating blade, large, exstipulate. <u>Flowers</u> solitary, axillary, actinomorphic, perfect, hypogynous, perigynous to sometimes epigynous, large and showy, floating or raised above the water surface, with long peduncles. <u>Calyx</u> of generally 4 or 6 sepals, distinct, sometimes petaloid, imbricate. <u>Corolla</u> of 8 to many petals in several whorls, staminodial in origin, distinct, often not clearly differentiated from the stamens, with the inner whorl(s) transitional or with all petals stamen-like, sometimes with abaxial nectaries, yellow, white, red, or blue, imbricate. <u>Androecium</u> of numerous laminar and spirally arranged stamens; filaments poorly differentiated from the anthers, distinct; anthers dehiscing longitudinally, introrse; sometimes inner whorl staminodial. <u>Gynoecium</u> of generally 1 pistil (with degree of carpel connation various), 5- to many-carpellate; ovary superior to sometimes inferior, with as many locules as carpels; ovules numerous per locule, anatropous, pendulous, placentation parietal or lamellate; styles absent; stigma(s) large and discoid, with receptive area(s) over the entire surface or restricted to radiating lines. <u>Fruit</u> berry-like, spongy, irregularly dehiscent (due to swelling of mucilage surrounding seeds); seeds small, usually operculate, often arillate; endosperm scanty; perisperm copious; embryo straight.

Family characterization in Florida: perennial aquatic herbs with large rhizomes and milky sap; large, long-petiolate leaves with floating, cordate to orbicular (peltate) blades; large, long-peduncled, solitary flowers with numerous petals, stamens, and carpels; laminar stamens undifferentiated into filament and anther; expanded discoid and

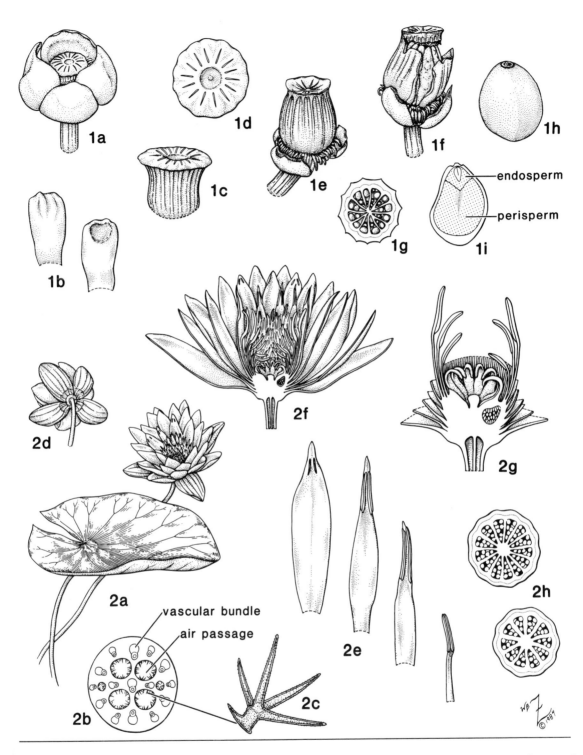

Figure 2. **Nymphaeaceae.** 1, *Nuphar luteum:* **a,** flower, x 1/2; **b,** two views of petal, x 2 1/4; **c,** pistil, x 1; **d,** top view of pistil showing radiating stigmatic lines, x 1; **e,** fruit, x 1/2; **f,** dehiscing fruit, x 2/5; **g,** cross section of fruit, x 1/2; **h,** seed, x 4; **i,** cross section of seed, x 4.   2, *Nymphaea odorata:* **a,** habit, x 1/4; **b,** cross section of petiole, x 6; **c,** branched sclereid, x 80; **d,** view of flower showing the 4 sepals, x 1/4; **e,** sequence showing transition from petaloid staminode of the outer whorl (top) to stamen of the inner whorl (bottom), x 1 1/2; **f,** longitudinal section of flower, x 1; **g,** longitudinal section of flower (perianth removed), x 2; **h,** cross section of ovaries from two different flowers, x 2.

CHART OF MAJOR MORPHOLOGICAL DIFFERENCES

Character	NYMPHAEACEAE	NELUMBONACEAE	CABOMBACEAE
Gen./spp. Fla. reps.	6/62 2/8	1/2 1/2	2/8 2/2
Habit	no vessels scapose	vessels in roots scapose	no vessels caulescent
Leaves	alternate, long-petiolate, cordate to peltate	alternate, long-petiolate, peltate	alternate, long-petiolate, cordate to peltate OR opposite or whorled with short petioles and dissected blades
Perianth	4 to 6 sepals (sometimes petaloid) + 8 to numerous petals (often not clearly differentiated from stamens)	2 "sepals" + numerous petals	3 + 3 tepals
Stamens	numerous laminar	numerous filament + anther with prolonged connective	3 to numerous filament + anther
Pollen	monocolpate	tricolpate	usually monocolpate
Carpels	5 to numerous basically fused	numerous distinct, individually embedded by enlarged, obconical receptacle	2 to 18 distinct
Ovules per carpel or locule	numerous	1	usually 2 or 3
Stigma	enlarged, discoid, and radiating	capitate	terminal or decurrent
Fruit	berry-like, spongy, irregularly dehiscent	aggregate of nuts, each free in cavities of enlarged receptacle	achene-like (nutlets)
Seeds	operculate often arillate scanty endosperm; copious perisperm	----- no arils essentially no endosperm; no perisperm	operculate no arils scanty endosperm copious perisperm
Embryo	small	large (fleshy cotyledons)	small

radiating stigma; spongy, dehiscent, berry-like fruit; and operculate seeds with perisperm. Pollen grains monocolpate. Anatomical features: parenchymatous tissues with schizogenous intercellular spaces and articulated laticifers; branched sclereids (especially in the leaves); scattered vascular bundles (in the axes, petioles, and peduncles) with no vessels or cambium; and root hairs that arise from specialized cells (as in many monocots).

Genera/species: 6/62

Distribution: Widespread from tropical to northern cold temperate regions in quiet freshwaters (e.g., ponds, streams, and lakes).

Major genera: *Nymphaea* (35 spp.) and *Nuphar* (19 spp.)

Florida representatives: *Nymphaea* (7 or 8 spp.) and *Nuphar* (1 sp.); see Godfrey and Wooten, (1981).

Economic plants and products: Edible seeds from species of *Victoria* and *Nymphaea* (rhizomes also edible). Ornamental plants for pools and aquaria: *Euryale* (prickly water-lily), *Nuphar* (yellow water-lily, spatter-dock), *Nymphaea* (water-lily), and *Victoria* (royal or Amazon water-lily).

The Nymphaeaceae are treated in a rather restricted sense here (see Thorne, 1974, 1983 and Cronquist, 1981); several authors also include species here referred to the Cabombaceae and Nelumbonaceae (see Wood, 1959 and Simon, 1971). These families share such morphological features as: the aquatic perennial habit; long-petiolate, cordate to peltate leaves; solitary, long-peduncled flowers; and parietal placentation, as well as

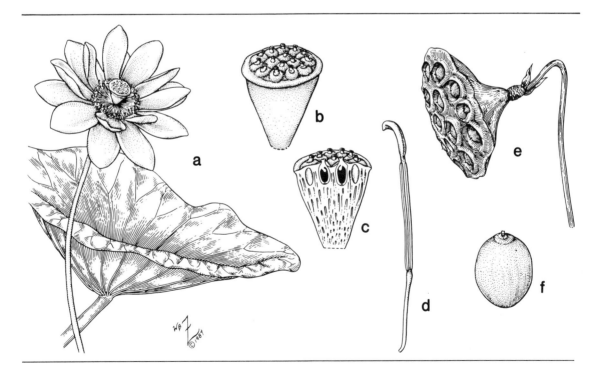

Figure 3. Nelumbonaceae. *Nelumbo lutea:* **a,** flower and emergent leaf, x 1/4; **b,** receptacle of flower (note sunken carpels), x 1; **c,** longitudinal section of receptacle, x 1; **d,** stamen, x 2; **e,** fruiting receptacle, x 2/5; **f,** nutlet, x 1 1/2.

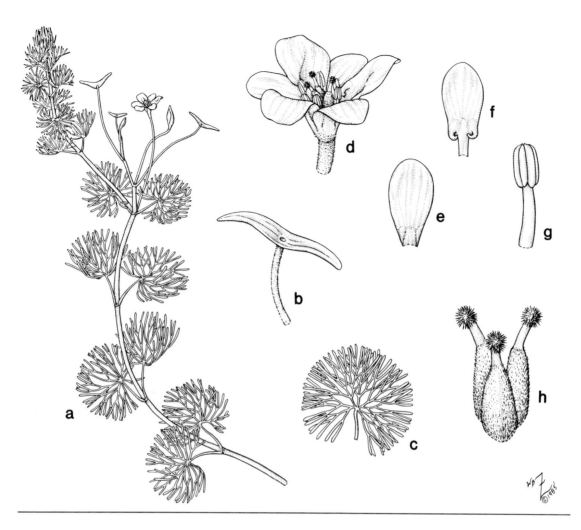

Figure 4. Cabombaceae. *Cabomba caroliniana:* **a,** habit, x 1/2; **b,** floating leaf, x 2 1/2; **c,** submersed leaf, x 2/3; **d,** flower, x 4; **e,** outer tepal, x 3; **f,** inner tepal, x 3; **g,** stamen, x 9; **h,** gynoecium, x 8.

several anatomical features. However, numerous characters, especially gynoecial features, have been used to separate these three groups (see chart and Figs. 2, 3, and 4). All three families are usually considered to be allies of the primitive families of dicots (Li, 1955).

The Nymphaeaceae have several distinctive anatomical characteristics that link them to the monocots. The plants lack a cambium and true vessels and have long tracheids with spiral or annular thickenings. As in the monocots, the vascular bundles are scattered within the ground tissue, and the root hairs originate from specialized cells.

The flowers of *Nuphar* and *Nymphaea* are distinctive with numerous petals and stamens and a conspicuous stigma (Moseley, 1961, 1965). The petals blend into the laminar stamens with transitional forms. In *Nuphar*, the petals are scale-like, and each has a nectary on the abaxial side. The receptive stigmatic surface is represented by radiating lines on the discoid stigma. The carpels of *Nymphaea* form a broad concave stigma, with each carpel having an incurved prolongation (see Fig. 2, 2g). In this case, the entire upper surface of each carpel is receptive.

The flowers are protogynous and are visited mainly by pollen-collecting insects (flies, beetles), which promote cross-pollination. Self-pollination may also occur in some species as the flowers open. Although the flowers of *Nymphaea* do not produce nectar, the receptive stigmatic cup fills up with a sweet liquid secreted by the stigma, and some species are also fragrant.

Water-lily seeds and fruits, which generally mature underwater, are adapted for water dispersal. The ripe fruit eventually bursts due to the swelling of the mucilage surrounding the seeds. After the outer rind of a *Nuphar* fruit ruptures, each protruding carpel separates (like a segment of an orange) and floats away. The released seeds of *Nymphaea* rise to the surface due to the air enclosed in their arils.

References Cited

Cronquist, A. 1981. An integrated system of classification of flowering plants. Pp. 100-115. Columbia Univ. Press, New York, NY.

Li, H.-L. 1955. Classification and phylogeny of Nymphaeaceae and allied families. Amer. Midl. Naturalist 54: 33-41.

Godfrey, R.K. and J.W. Wooten. 1981. Aquatic and wetland plants of the southeastern United States. Dicotyledons. Pp. 160-170. Univ. of Georgia Press, Athens, GA.

Moseley, M.F. 1961. Morphological studies of the Nymphaeaceae. II. The flower of *Nymphaea*. Bot. Gaz. (Crawfordsville) 122: 233-259.

_____. 1965. Ibid. III. The floral anatomy of *Nuphar*. Phytomorphology 15: 54-84.

Simon, J.-P. 1971. Comparative serology of the order Nymphaeales. II. Relationships of Nymphaeaceae and Nelumbonaceae. Aliso 7: 325-350.

Thorne, R.F. 1974. A phylogenetic classification of the Annoniflorae. Aliso 8: 147-209.

_____. 1983. Proposed new realignments in the angiosperms. Nord. J. Bot. 3: 85-117.

Wood, C.E. 1959. The genera of Nymphaeaceae and Certophyllaceae in the southeastern United States. J. Arnold Arbor. 40: 94-112.

SARRACENIACEAE (PITCHER-PLANT FAMILY)

Perennial herbs growing in marshes or bogs, insectivorous, with rhizomes. <u>Leaves</u> simple, alternate, in basal rosettes, highly modified, tubular or trumpet-shaped with ridge or laminar wing on adaxial side and a terminally expanded hood, often with retrorse hairs on the inner surface, liquid-filled at base, green, yellow, maroon, or variegated, exstipulate. <u>Flowers</u> solitary and terminal actinomorphic, perfect, hypogynous, large, nodding. <u>Calyx</u> of usually 5 sepals, distinct, often colored and petaloid, imbricate, persistent. <u>Corolla</u> of usually 5 petals, distinct, caducous, usually yellow, maroon, or red, imbricate. <u>Androecium</u> of numerous stamens; filaments distinct; anthers basifixed or versatile, dehiscing longitudinally, extrorse. <u>Gynoecium</u> of 1 pistil, usually 5-carpellate; ovary superior, usually 5-loculed and -lobed; ovules numerous, anatropous, placentation basically axile (with partitions above often not meeting or joined); style 1, usually with greatly expanded peltate apex, with 5 lobes, persistent; stigmas 5, small and restricted to area beneath each style lobe tip. <u>Fruit</u> a loculicidal capsule; seeds numerous, small, often winged; endosperm copious, oily, firm-fleshy; embryo minute, linear.

Family characterization in Florida: perennial insectivorous herbs growing in marshes or bogs; highly modified and specialized tubular leaves in basal rosettes; solitary, nodding, perfect flowers with 5-merous perianth and gynoecium; numerous stamens; and dilated peltate style with a restricted stigmatic area under each of 5 lobes. Several

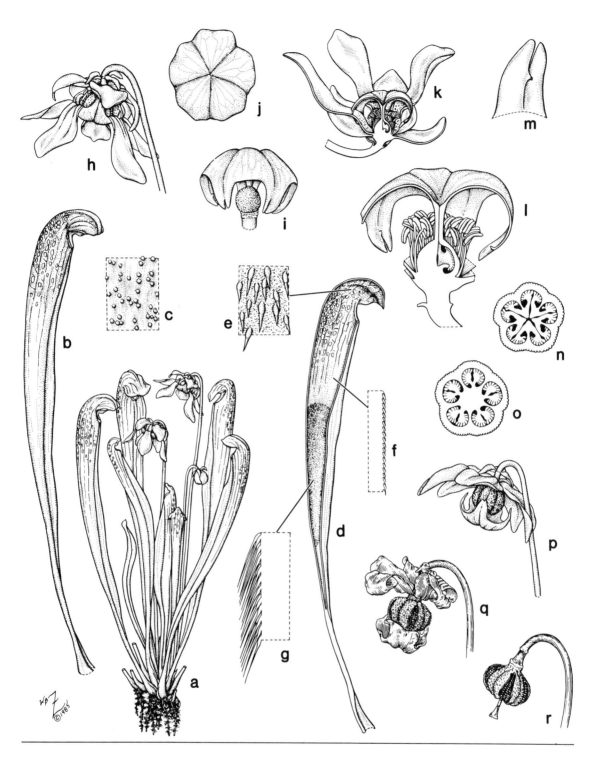

Figure 5. Sarraceniaceae. *Sarracenia minor:* **a**, habit, x 1/5; **b**, leaf, x 1/3; **c**, glands on outer surface of leaf, x 25; **d**, longitudinal section of leaf, x 1/3; **e**, glandular hairs along inner surface of hood, x 25; **f**, detail of upper section of leaf ("smooth zone"), x 25; **g**, detail of lower section of leaf, x 25; **h**, flower, x 1/2; **i**, pistil, x 1; **j**, top view of expanded style, x 1; **k**, longitudinal section of flower, x 2/3; **l**, longitudinal section of flower (perianth removed), x 1 3/4; **m**, tip of style branch with stigma, x 3; **n**, cross section of ovary near middle, x 3 1/2; **o**, cross section of ovary at base, x 3 1/2; **p**, young fruit, x 1/2; **q**, capsule with persistent calyx, x 2/3; **r**, capsule (calyx removed), x 2/3.

anatomical features related to specialized leaves (see Metcalfe and Chalk, 1950 and discussion below).

Genera/species: 3/15

Distribution: In marshy and boggy habitats of eastern North America, coastal Oregon and northern California, and northern South America.

Major genus: *Sarracenia* (10 spp.)

Florida representatives: *Sarracenia* (6 spp.); see McDaniel (1966, 1971).

Economic plants and products: Novelty house plants: *Sarracenia* (pitcher-plant) and *Darlingtonia* (cobra-plant, California pitcher-plant).

The Sarraceniaceae, a distinctive group, have been allied with various families (see Wood, 1960; DeBuhr, 1975). The morphology of the leaves ("pitchers") provides the most important taxonomic characters used in distinguishing species.

The tubular leaves of *Sarracenia* consist of gracefully curved pitchers, each covered by a hood or lid that represents a prolongation of the dorsal side. A ridge or laminar wing is present along the ventral surface. Insects are attracted to the bright coloring and often strong odors of the leaves. Several species also have bright, window-like perforations (fenestrations) around the necks of the pitchers. In addition, the outer surface of the leaf and inner surface of the lid and mouth generally have secreting glandular hairs. After entering the pitcher, an insect may slip downward into the leaf along a slick zone composed of epidermal cells with downwardly directed projections (covering the surface like fish scales). It next encounters long, recurved hairs (which cover most of the tube base) and becomes trapped within the leaf, later to drown in the liquid accumulated at the base. In most species of *Sarracenia*, the prey is digested by acids and enzymes and then absorbed by the plant. Despite the insectivorous nature of the leaves, several insect larvae (of certain flies and mosquitoes) may inhabit the pitchers (Lloyd, 1942; Slack, 1980).

Many flowers have a musty or sometimes sweet odor; otherwise little has been reported about pollination mechanisms.

Hybridization occurs within *Sarracenia*, although species that occur in the same area tend to be temporally isolated by flowering periods.

References Cited

DeBuhr, L.E. 1975. Phylogenetic relationships of the Sarraceniaceae. Taxon 24: 297-306.

Lloyd, F.E. 1942. The carnivorous plants. Pp. 17-39. Chronica Botanica, Waltham, MA.

McDaniel, S.T. 1966. A taxonomic revision of *Sarracenia* (Sarraceniaceae). 128 pp. Ph.D. Dissertation. Florida State Univ., Tallahassee, FL.

_____. 1971. The genus *Sarracenia* (Sarraceniaceae). Bull. Tall Timbers Research Station 9: 1-36.

Metcalfe, C.R. and L. Chalk. 1950. Anatomy of the dicotyledons. Vol. 2. Pp. 71-74. Oxford Univ. Press, Oxford, England.

Slack, A. 1980. Carnivorous plants. Pp. 25-69. The M.I.T. Press, Cambridge, MA.

Wood, C.E. 1960. The genera of Sarraceniaceae and Droseraceae in the southeastern United States. J. Arnold Arbor. 41: 152-163.

SAPOTACEAE (SAPODILLA OR SAPOTE FAMILY)

Trees or shrubs, with milky sap. <u>Leaves</u> simple, entire, alternate, sometimes pseudoverticillate, coriaceous, usually exstipulate. <u>Inflorescence</u> determinate, cymose with flowers often in fascicles and appearing umbellate, or sometimes flower solitary, axillary, sometimes cauliflorous and occurring at nodes on old wood. <u>Flowers</u> actinomorphic, perfect, hypogynous, small. <u>Calyx</u> of often 5 (uniseriate) or 4, 6, or 8 (biseriate) sepals, distinct or basally connate, imbricate. <u>Corolla</u> sympetalous, usually with as many lobes as sepals, sometimes with paired petaloid appendages, often white or cream-colored, imbricate. <u>Androecium</u> of basically 8 to 15 stamens in 2 or 3 whorls of 4 or 5, usually with the outer whorl(s) reduced to staminodes (often petaloid) or absent, epipetalous; filaments distinct; anthers basifixed, dehiscing longitudinally, extrorse. <u>Gynoecium</u> of 1 pistil, usually 4- or 5-carpellate; ovary superior, usually 4- or 5-loculed, usually hirsute; ovules 1 in each locule, anatropous (to hemitropous), placentation axile or axile-basal; style 1; stigma 1, capitate or slightly lobed, inconspicuous. <u>Fruit</u> a berry, often with a thin, leathery to bony outer layer; seeds 1 to few, large, often with thick and hard seed coat and conspicuous hilum (scar); endosperm scanty to copious; oily, fleshy, or hard, or absent; embryo large.

Family characterization in Florida: trees or shrubs with laticiferous parts; thick, sympetalous corolla, each lobe often with paired appendages; 2 or 3 whorls of epipetalous stamens and petaloid staminodes; a berry as the fruit type; and seeds with hard and thick testa and large hilum. Tissues commonly with calcium oxalate crystals and tannins. Anatomical features: well-developed latex-sacs (in leaves, bark, and pith); vestiture of unicellular 2-armed hairs ("malpighian hairs"; sometimes one branch suppressed); three-trace trilacunar nodes; and unitegmic and tenuinucellar ovules.

Genera/species: 35-75/800

Distribution: Pantropical; especially in lowland and montane rain forests.

Major genera: *Pouteria* (150 spp.), *Chrysophyllum* (90-150 spp.), *Palaquium* (115+ spp.), and *Planchonella* (100 spp.)

Florida representatives: 6 genera/12 spp.; largest genera: *Bumelia* (5 spp.), *Manilkara* (2 spp.), and *Pouteria* (2 spp.)

Economic plants and products: Chicle (essential chewing gum ingredient) from *Manilkara* (= *Achras*). Gutta-percha (rubbery compounds) from several, such as species of *Mimusops*, *Palaquium*, and *Payena*. Edible fruits from: *Chrysophyllum* (star-apple), *Manilkara* (sapodilla), *Pouteria* (eggfruit or canistel), and *Calocarpum* (mamey sapote). Edible oils (from seeds) of *Butyrospermum* (shea-butter) and *Madhuca*. Very durable timber from: *Manilkara* (bulletwood), *Mimusops* (cherry-mahogany), and *Sideroxylon* (iron-wood). Ornamental plants (species of 10 genera), including: *Bumelia* (buckthorn), *Chrysophyllum* (satin-leaf), *Dipholis*, *Manilkara*, and *Pouteria*.

The delimitation of the genera (35-75) and species (600-800) of the Sapotaceae is extremely controversial (see Metcalfe and Chalk, 1950; Wood and Channell, 1960; and Cronquist, 1981), with various authors splitting and/or combining groups, resulting in confusing nomenclature. Seed characters (presence of endosperm, morphology of hilum) and the presence of staminodia and/or of appendages on the corolla have been important features for separating genera (Cronquist, 1946a,b).

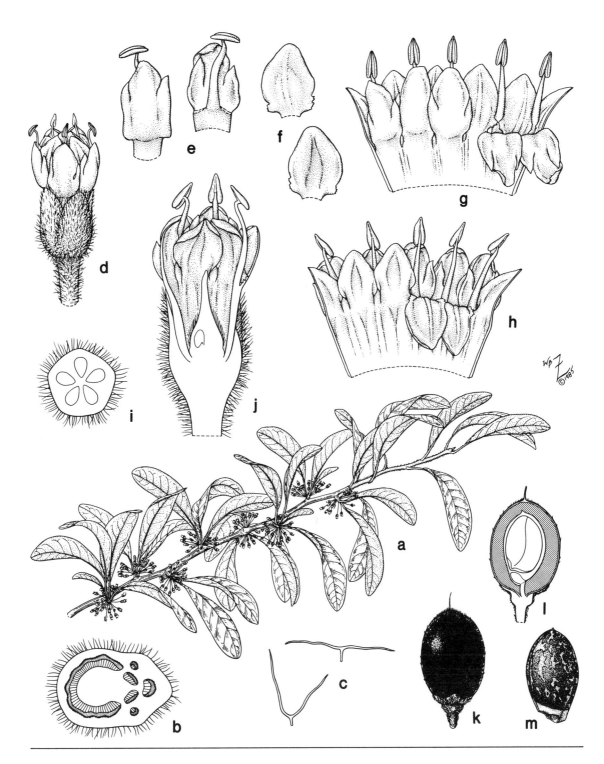

Figure 6. Sapotaceae. *Bumelia lanuginosa:* **a,** flowering branch, x 1/3; **b,** cross section of node (three-trace trilacunar), x 12; **c,** unicellular two-armed hairs ("malpighian hairs") from abaxial surface of leaf, x 25; **d,** flower, x 7; **e,** two views of corolla lobe and adnate stamen, x 12; **f,** two views of staminode, x 12; **g,** outer (abaxial) surface of expanded corolla and androecium with two corolla lobes folded down, x 10; **h,** inner (adaxial) surface of expanded corolla and androecium with two staminodia folded down, x 10; **i,** cross section of ovary, x 25; **j,** longitudinal section of flower, x 12; **k,** berry, x 3; **l,** longitudinal section of berry, x 3; **m,** seed, x 3 1/2.

The perianth is extremely variable within the family. For example, the sepals vary in number and arrangement: five and uniseriate (*Bumelia*) or biseriate with two (*Pouteria*), three (*Manilkara*), or four sepals in each whorl. The number of corolla lobes is usually equal to that of the sepals. In addition, paired petaloid appendages may occur on each corolla lobe, as well as petaloid staminodes that alternate with the lobes.

Pollination has not been thoroughly examined. In several (e.g., *Bumelia celastrina* and *Dipholis salicifolia*), the receptive stigma protrudes before the flower opens and before the anthers dehisce (Tomlinson, 1980). In others (*B. reclinata*), the style remains enclosed by the corolla appendages, which diverge later to expose the stigma after the pollen is shed. Flowers in some groups (such as *Chrysophyllum* species) are sweet-scented. Many flowers are nocturnal and bat-pollinated.

References Cited

Cronquist, A. 1946a. Studies in the Sapotaceae -- II. Survey of the North American genera. Lloydia 9: 241-292.

_____. 1946b. Ibid. VI. Miscellaneous notes. Bull. Torrey Bot. Club 73: 465-471.

_____. 1981. An integrated system of classification of flowering plants. Pp. 496-499. Columbia Univ. Press, New York, NY.

Metcalfe, C.R. and L. Chalk. 1950. Anatomy of the dicotyledons. Vol 2. Pp. 871-880. Oxford Univ. Press, Oxford, England.

Tomlinson, P.B. 1980. The biology of trees native to tropical Florida. Pp. 382-396. "Publ. by the author", Petersham, MA.

Wood, C.E. and R.B. Channell. 1960. The genera of Ebenales in the southeastern United States. J. Arnold Arbor. 41: 1-35.

CARYOPHYLLACEAE (PINK FAMILY)

Usually annual or perennial herbs; stems typically with swollen nodes. Leaves simple, entire, opposite, decussate, usually narrow, appearing parallel veined, often basally connate or connected by a transverse line, usually exstipulate. Inflorescence determinate, basically cymose, sometimes flower solitary, usually terminal. Flowers actinomorphic, usually perfect, usually hypogynous. Calyx of 5 or occasionally 4 sepals, distinct to connate, imbricate, often with membranous margins, persistent. Corolla of 5 or occasionally 4 petals, sometimes reduced or absent, distinct, often differentiated into claw and limb with appendages on the inner surface of the claw-limb junction, often apically notched, commonly white or pink, imbricate. Androecium of 5 (uniseriate) to 10 (biseriate with outer whorl apparently opposite the petals) or occasionally 1 to 4 stamens; filaments distinct, sometimes basally adnate to the petals or calyx, often with nectaries at base; anthers basifixed, dehiscing longitudinally. Gynoecium of 1 pistil, 2- to 5-carpellate; ovary superior, usually 1-loculed (at least above), sometimes 3- to 5-loculed at the base, often stalked; ovules usually numerous, sometimes solitary to few, usually campylotropous, placentation free central (in 1-loculed ovary), free central above and axile below (in ovary partitioned at base) or occasionally basal (when ovule solitary); styles 2 to 5; stigmas 2 to 5, minute. Fruit usually a capsule dehiscing apically by valves or teeth, or sometimes a utricle; perisperm hard; embryo curved, peripheral.

Family characterization in Florida: herbaceous plants with swollen nodes; simple, opposite, and narrow leaves with entire margins and connate or connected bases; pink or white petals (when present) often with an apical notch; 2- to 5-carpellate unilocular ovary with free central placentation (at least above); a capsule (dehiscing by apical valves or teeth) or utricle as the fruit type; and seeds with perisperm and a peripheral, curved

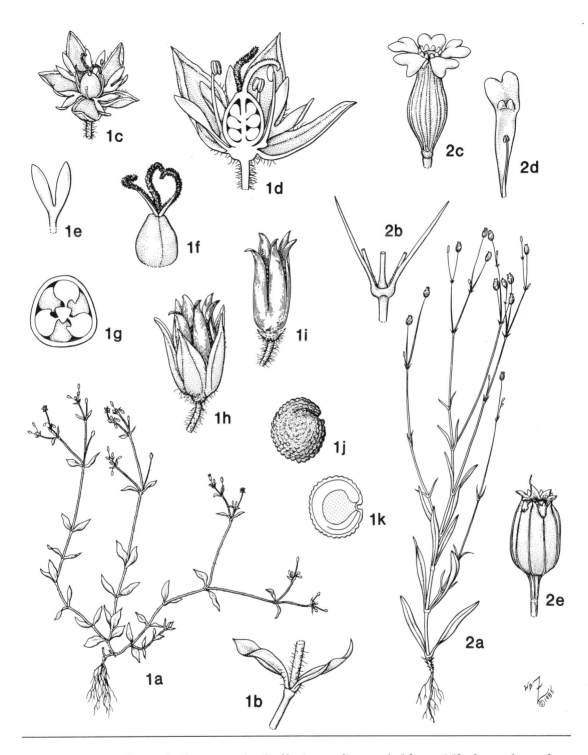

Figure 7. Caryophyllaceae. 1, *Stellaria media:* **a,** habit, x 1/3; **b,** node, x 3; **c,** flower, x 6; **d,** longitudinal section of flower, x 12; **e,** petal, x 12; **f,** pistil, x 12; **g,** cross section of ovary, x 22; **h,** capsule with persistent calyx, x 6; **i,** capsule (calyx removed), x 6; **j,** seed, x 15; **k,** longitudinal section of seed, x 15. **2,** *Silene antirrhina:* **a,** habit, x 1/2; **b,** node, x 2; **c,** flower, x 4 1/2; **d,** petal and adnate stamen, x 6; **e,** capsule, x 4 1/2.

embryo. Saponins (glucosides) and lychnose (unusual storage carbohydrate) present. Anatomical feature: sieve tubes with a special kind of P-type plastid (Behnke, 1976).

Genera/species: 70/1,750

Distribution: Primarily in north temperate regions with a few representatives in south temperate zones, montane tropics, and the arctic; especially diverse in the Mediterranean region.

Major genera: *Silene* (400-500 spp.), *Dianthus* (300 spp.), *Arenaria* (250 spp.), and *Gypsophila* (125 spp.)

Florida representatives: 14 genera/27 spp.; largest genera: *Paronychia* (7 spp.), *Arenaria* (4 spp.), *Silene* (3 spp.), *Cerastium* (2 spp.), and *Stellaria* (2 spp.)

Economic plants and products: Many widespread weedy plants, such as *Cerastium* and *Stellaria* (chickweeds). Garden ornamentals (species of 22 genera), including: *Arenaria* (sandwort), *Cerastium* (mouse-ear chickweed), *Dianthus* (carnation, pink, sweet William), *Gypsophila* (baby's-breath), *Lychnis* (Maltese-cross), *Saponaria* (soapwort), and *Silene* (catchfly).

The Caryophyllaceae are usually divided into three distinct subfamilies based primarily upon the presence of stipules, fusion of the sepals, and the morphology of the petals. The phylogenetic relationships of the family have been closely scrutinized since the Caryophyllaceae differ from most allied families (in the Centrospermae or Chenopodiiflorae) in having anthocyanins (as in most flowering plants) and not the betalains found in these groups (Eckardt, 1976; Mabry et al., 1963, 1972; Mabry, 1974, 1977). Although a uniform and easily recognizable family, the generic limits are somewhat difficult and controversial.

The flowers of the family demonstrate a range of morphological complexity in such characters as fusion of the calyx, presence and morphology of the corolla, and number of stamens and carpels. Variation may even occur in the same species, as for example, *Stellaria media* with five, ten, or three stamens. The petals of the Caryophyllaceae are generally considered to be derived from the stamens (Thomson, 1942; Cronquist, 1981) and are often either bifid at the apex or clawed with appendages ("corona") at the claw-limb junction.

Pollinators visit for the nectar secreted at the base of the stamens. Flies and bees generally pollinate the more open type flowers, such as *Stellaria*. Others in the family (*Silene*) conceal the nectar in the tube formed by the synsepalous calyx and long-clawed and appendaged petals, and these flowers are probed by larger bees and Lepidoptera. The flowers are commonly protandrous, although self-pollination occurs in several species.

The unusual capsule, which opens by recurving, apical teeth, is characteristic of most species. Wind or animals are required to shake the capsules in order to disperse the seeds. In a few otherw (such as *Paronychia*), the fruit is a utricle.

References Cited

Behnke, H.-D. 1976. Ultrastructure of sieve-element plastids in Caryophyllales (Centrospermae), evidence for delimitation and classification of the order. Pl. Syst. Evol. 126: 31-54.

Cronquist, A. 1981. An integrated system of classification of flowering plants. Pp. 272-276. Columbia Univ. Press, New York, NY.

Eckardt, Th. 1976. Classical morphological features of centrospermous families. Pl. Syst. Evol. 126: 5-25.

Mabry, T.J. 1974. Is the order Centrospermae monophyletic? Pp. 275-285 *in:* G. Bendz and J. Santesson (eds.), Chemistry in botanical classification, Nobel symposium 25. Academic Press, Inc., London, England.

_____. 1977. The order Centrospermae. Ann. Missouri Bot. Gard. 64: 210-220.

_____, A. Taylor, and B.L. Turner. 1963. The betacyanins and their distribution. Phytochemistry 2: 61-64.

_____, L. Kimler, and C. Chang. 1972. The betalains: structure, function, and biogenesis, and the plant order Centrospermae. Pp. 105-134 *in:* V.C. Runeckles and T.C. Tso (eds.), Recent advances in phytochemistry. Academic Press, Inc., London, England.

Thomson, B.F. 1942. The floral morphology of the Caryophyllaceae. Amer. J. Bot. 29: 333-349.

CHENOPODIACEAE (GOOSEFOOT FAMILY)

Annual or perennial herbs or sometimes shrubs, generally growing in disturbed, saline or xeric habitats, often succulent. <u>Leaves</u> simple, entire or sometimes toothed or lobed, usually alternate, generally succulent, sometimes terete or reduced to scales, often covered with hairs (causing a "mealy" appearance), exstipulate. <u>Inflorescence</u> determinate, basically cymose (with flowers congested in leaf axils) and often appearing spicate, racemose, or paniculate, axillary. <u>Flowers</u> usually actinomorphic, perfect or less often imperfect (then plants dioecious to monoecious), hypogynous, minute, bracteate. <u>Perianth</u> of usually 5 tepals, uniseriate, distinct or basally connate, herbaceous to membranous, green or greenish, more or less imbricate, persistent, generally accrescent. <u>Androecium</u> of usually 5 stamens (opposite the tepals), arising from receptacle, inserted on disc, or adnate to perianth; filaments usually distinct, incurved in bud; anthers basifixed, dehiscing longitudinally, introrse or latrorse. <u>Gynoecium</u> of 1 pistil, 2- or 3-carpellate; ovary superior, 1-loculed; ovule solitary, campylotropous or sometimes amphitropous, placentation basal; style usually 1; stigmas 2 or 3, filiform. <u>Fruit</u> an achene, utricle, or occasionally a pyxis, often subtended by persistent tepals and bracts, sometimes forming a multiple fruit by connation of tepals of several flowers; seeds lenticular; perisperm copious, starchy, usually hard; embryo curved or spirally twisted, surrounding the perisperm.

Family characterization in Florida: more or less fleshy herbs to shrubs growing in weedy, xeric, or saline habitats; minute flowers in dense cymose inflorescences; greenish, herbaceous to membranous, uniseriate perianth of 5 tepals; 5 distinct stamens; unilocular ovary with a solitary basal ovule; and perisperm surrounded by curved or coiled embryo. Betalains (nitrogen-containing pigments) present (see Amaranthaceae in *Part I*). Tissues with calcium oxalate crystals. Anatomical features: anomalous secondary growth (similar to that of the Amaranthaceae; see *Part I*) and sieve tubes with a special kind of P-type plastid (Behnke, 1976).

Genera/species: 102/1,400

Distribution: Cosmopolitan; especially abundant in weedy, xeric or saline areas.

Major genera: *Chenopodium* (at least 200 spp.), *Atriplex* (200 spp.), *Obione* (100+ spp.), *Salsola* (150 spp.), and *Suaeda* (110 spp.)

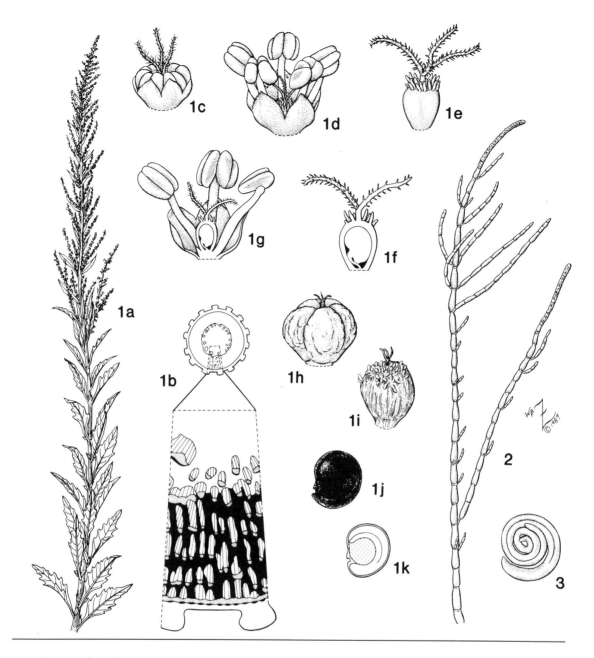

Figure 8. Chenopodiaceae. 1, *Chenopodium ambrosioides:* **a,** habit, x 1/3; **b,** cross section of stem, x 2, with detail showing anomalous secondary growth pattern, x 15; **c,** flower before extension of stamens, x 12; **d,** flower, x 12; **e,** pistil, x 25; **f,** longitudinal section of pistil, x 25; **g,** longitudinal section of flower, x 15; **h,** fruit (utricle subtended by accrescent perianth), x 18; **i,** utricle, x 25; **j,** seed, x 25; **k,** longitudinal section of seed, x 25. **2,** *Salicornia virginica:* habit, x 1/2. **3,** *Suaeda maritima:* embryo, x 12.

Florida representatives: *Chenopodium* (8 spp.), *Atriplex* (2 spp.), *Salicornia* (2 spp.), *Suaeda* (2 spp.), and *Salsola* (1 sp.)

Economic plants and products: Food plants: *Beta vulgaris* (beet, Swiss chard), *Spinacia oleracea* (spinach), and several species of *Chenopodium* (edible greens, pseudo-grains from mealy perisperm of seeds). Several weedy plants, such as species of *Chenopodium* (goosefoot, lamb's-quarters) and *Salsola* (Russian-thistle). Ornamental plants (species of 7 genera), including: *Atriplex* (saltbush), *Chenopodium*, *Kochia* (summer-cypress), and *Salicornia* (glasswort).

The Chenopodiaceae are usually divided into two or three subfamilies (see Williams and Ford-Lloyd, 1974 and Blackwell, 1977) based on characters of the embryo (curved or spirally coiled).

The varied vegetative habits within the family include adaptations for habitats with soils containing a large percentage of inorganic salts. The species of seashores (and similar habitats) are generally succulent and brittle plants with reduced (*Salsola*) to nearly absent (*Salicornia*) leaves. Others in the family, such as *Chenopodium*, are leafy weeds inhabiting salt rich soils around dwellings and disturbed areas. Species of such varied genera as *Chenopodium* and *Salsola* often have a vesture of hairs with distended, thin-walled apices containing water and oxalates. The desiccated hairs appear as white flakes on mature plant parts, causing a "mealy" appearance. Water is also retained in water-storage tissue often present in the mesophyll.

The inconspicuous flowers are rarely visited by insects and are generally anemophilous. Nectariferous glands or a disc may occur at the filament bases of perfect flowers, but imperfect flowers lack nectaries. Protandry or protogyny promotes cross-pollination, although self-pollination does occur in several species.

References Cited

Behnke, H.D-. 1976. Ultrastructure of sieve-element plastids in Caryophyllales (Centrospermae), evidence for delimitation and classification in the order. Pl. Syst. Evol. 126: 31-54.

Blackwell, W.H. 1977. The subfamilies of the Chenopodiaceae. Taxon 26: 395-397.

Williams, J.T. and B.V. Ford-Lloyd. 1974. The systematics of the Chenopodiaceae. Taxon 23: 353-354.

CACTACEAE (CACTUS FAMILY)

Perennial herbs, vines, shrubs, or sometimes small trees, often growing in xeric habitats, sometimes epiphytic, succulent, with leaves generally reduced to spines or quickly deciduous, with watery or mucilaginous sap; stems usually greatly enlarged and cylindrical, conical, or flattened, often tubercled or ribbed, simple or branched, often jointed; roots usually superficial, slender, fleshy. Leaves simple, alternate, fleshy, sometimes large and persistent to usually rudimentary (scale-like) and caducous, with axillary buds (or branches) specialized into cushion-like areas (areoles) bearing spine clusters (modified axillary shoot leaves) and sometimes bristles (glochids). Flowers usually solitary, borne upon or near the areoles, actinomorphic to sometimes slightly zygomorphic, perfect, epigynous, often large and showy, sessile, with nectariferous ring along inner surface of hypanthium. Perianth of numerous intergrading sepaloid to petaloid tepals, spirally arranged, basally connate (forming a hypanthium), generally red, purple, orange, yellow, or white. Androecium of numerous stamens, spirally arranged or clustered, arising from hypanthium; filaments distinct; anthers basifixed or dorsifixed, dehiscing longitudinally, introrse or latrorse. Gynoecium of 1 pistil, 3- to many-carpellate; ovary inferior (see Boke, 1964), 1-loculed, usually embedded into the stem, commonly covered with hairs, bristles, or spines; ovules numerous, campylotropous to anatropous, with funiculi often connate at base and forming a bundle, placentation parietal; style 1; stigmas as many as carpels, usually radiating, thick, soft papillose. Fruit a berry, often spiny or bristly; seeds numerous, immersed in pulp; perisperm absent or sometimes abundant, starchy and mealy.

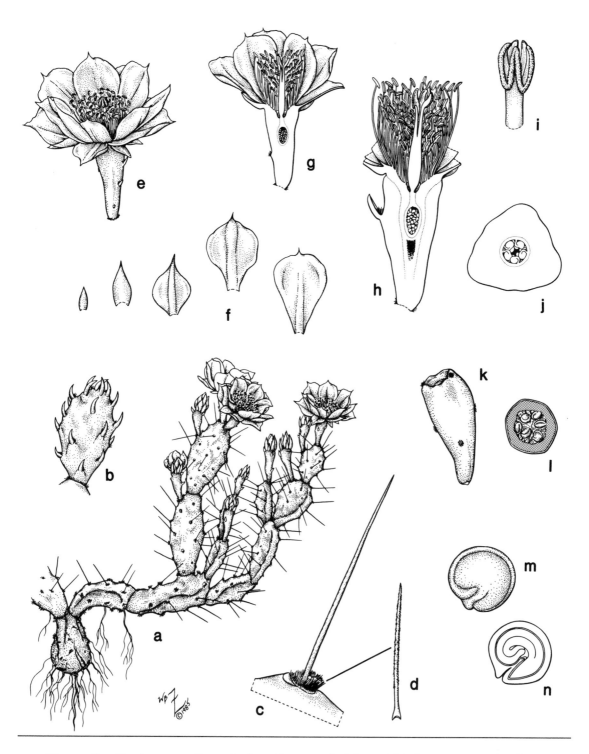

Figure 9. Cactaceae. *Opuntia humifusa:* **a,** habit, x 1/4; **b,** young pad with scale-like leaves, x 2/3; **c,** detail of areole, x 2; **d,** glochid, x 14; **e,** flower, x 2/3; **f,** sequence showing transition from sepaloid tepal of the outer whorl (left) to petaloid tepal of the inner whorl (right), x 2/3; **g,** longitudinal section of flower, x 2/3; **h,** longitudinal section of flower (perianth removed), x 1 1/3; **i,** stigma, x 3; **j,** cross section of ovary, x 2; **k,** berry, x 2/3; **l,** cross section of berry, x 3/4; **m,** seed, x 4 1/2; **n,** longitudinal section of seed, x 4 1/2.

Family characterization in Florida: succulent herbs to small trees usually growing in warm and dry habitats; enlarged stems that are often tubercled, ribbed and/or jointed; superficial roots; reduced to absent leaves with specialized axillary bud areas (areoles) bearing spines and sometimes bristles (glochids); solitary and showy flowers with intergrading sepals and petals; numerous spirally arranged tepals and stamens fused basally to form a hypanthium; 1-loculed, inferior ovary sunken into the stem; parietal placentation; and baccate, often spiny fruit. Betalains (nitrogen-containing pigments) present (see Amaranthaceae in *Part I*). Anatomical features: parenchymatous tissues commonly with scattered mucilage cells and calcium oxalate crystals; sieve tubes with a special kind of P-type plastid, and several adaptations of the stem related to water storage and retention (see discussion below).

Genera/species: 105/1,550

Distribution: Primarily localized in semi-desert areas of North, Central, and South America; a few genera (probably all introduced) in other areas of the world (e.g., Africa, Ceylon, India, and Australia).

Major genera: *Opuntia* (250 spp.) and *Mammillaria* (200-300 spp.).

Florida representatives: *Opuntia* (10 spp.), *Cereus* (7 or 8 spp.), *Pereskia* (2 spp.), and *Rhipsalis* (1 sp.); see Snow (1981).

Economic plants and products: Timber from several, such as species of *Pereskia* and *Cephalocereus*. Medicinal and ritual drugs from a few, such as *Lophophora* (peyote or mescal-button -- with mescaline, a narcotic alkaloid). Edible fruits from species of *Opuntia* (prickly-pear). Ornamental plants (species of at least 100 genera), including: *Cereus* (hedge cactus), *Echinopsis* (sea-urchin cactus), *Epiphyllum* (orchid cactus), *Hylocereus* (night-blooming cactus), *Lobivia* (cob cactus), *Mammillaria* (pincushion cactus), *Notocactus* (ball cactus), *Opuntia* (prickly-pear, cholla), *Rebutia* (crown cactus), *Rhipsalis* (mistletoe cactus), and *Schlumbergera* (Christmas cactus, crab cactus, Thanksgiving cactus).

The Cactaceae are generally divided into three subfamilies (or tribes) based on seed characters and the presence or absence of leaves, glochids, and flower stalks (see classifications in Britton and Rose, 1923 and Benson, 1982). Generic and specific delimitations have been difficult, due in part to the difficulty in preserving the succulent plants for comparative herbarium studies. Genera and species tend to be overdescribed resulting in a range of 30 to 300 genera and 1,000 to over 2,000 species cited in the literature. Collectors and commercial growers have probably also contributed to the proliferation of new "genera" and "species."

The cacti are mostly highly specialized stem-succulent herbs and rarely resemble the usual leaf-bearing type of plant. The enlarged stems, often with tubercles or prominent ribs, vary in shape from globose to cylindrical or flattened. Leaves, when present, are usually ephemeral. Except in a few taxa with well-developed leaves (*Pereskia*), photosynthesis (with "succulent metabolism" or CAM) takes place primarily in the stems (see Benson, 1982). Specializations of the stem structure for water storage and retention include the thick epidermis with sunken stomata and strong cuticle. Beneath this outer layer occurs a collenchymatous hypodermis surrounding a well-developed ground tissue of water-storage cells. The vascular system forms a cylindrical network embedded in the ground tissue.

The spines on the stem may also help promote the accumulation of dew or water droplets around the plant. Spines are restricted to hairy, cushion-like areas called areoles, which usually occur at the tips of the tubercles or along the ridges or edges of the stem. An areole is regarded as a modified axillary bud (or short branch), and a spine represents a specialized leaf (or bud scale) of that axillary branch. Besides protection from herbivores, the spines assist in vegetative propagation by animal dispersal. In some cacti (*Opuntia*), tufts of barbed bristles (glochids) also occur in the areoles.

The showy flowers of many species are often rapidly produced during or at the end of a rainy season, and they usually last only one or a few days. Several (*Cereus* and relatives) are night-blooming. Bees, beetles, birds, bats, and sphinx moths visit many species for the nectar secreted along the inner surface of the hypanthium and/or for the copious pollen. Some cacti (e.g., certain *Opuntia* species) have sensitive stamens that incurve when stimulated by crawling insects. With the anthers closer to the stigmas, the passageway for the pollinator becomes narrower. This does not necessarily promote self-pollination since most cacti are protandrous (and self-sterile). The stigmas are appressed when the anthers dehisce and later become erect or spread out over the stamens.

The baccate fruits are often sweet and tasty and dispersed by animals.

References Cited

Benson, N.L. 1982. The cacti of the United States and Canada. 1044 pp. Stanford Univ. Press, Stanford, CA.

Boke, N.H. 1964. The cactus gynoecium: a new interpretation. Amer. J. Bot. 51: 598-610.

Britton, N.L. and J.N. Rose. 1923. The Cactaceae. Vol. I. 236 pp. Vol II. 241 pp. Vol. III. 318 pp. Vol. IV. 258 pp. Publ. Carnegie Inst. Wash. No. 248.

Snow, B.L. 1981. A history of the Cactaceae of the southeastern United States. Cact. Succ. J. (Los Angeles) 53: 177-182.

OXALIDACEAE (OXALIS, SHEEP-SORREL, OR WOOD-SORREL FAMILY)

Perennial or annual herbs, sometimes suffrutescent or shrubs, with acrid juice, often with fleshy rhizomes, bulb-like tubers, and contractile roots. Leaves pinnately or more often palmately compound (then usually trifoliolate), or sometimes simple due to suppression of leaflets, alternate, often forming basal rosettes or apical clusters, with long petioles and leaflets often emarginate at apex, with pulvinus at petiole base and at each petiolule base, usually exstipulate. Inflorescence determinate, cymose and often appearing umbellate or sometimes racemose, axillary or seemingly terminal. Flowers actinomorphic, perfect, hypogynous, often heterostylous (dimorphic or trimorphic), sometimes cleistogamous. Calyx of 5 sepals, distinct or basally connate, imbricate, persistent. Corolla of 5 petals, distinct or sometimes basally connate, often clawed, often yellow, white, or purple, convolute or sometimes imbricate. Androecium of 10 stamens, biseriate with the outer whorl shorter than the inner whorl, opposite the petals, and sometimes reduced to staminodes; filaments basally connate and forming a ring or tube (monadelphous), those of the outer whorl with basal nectariferous thickening or glands, persistent; anthers dorsifixed, versatile, dehiscing longitudinally, introrse. Gynoecium of 1 pistil, 5-carpellate with carpels often incompletely connate (fused only along adaxial sutures); ovary superior, 5-loculed; ovules 1 or more in each locule, anatropous or sometimes hemitropous, pendulous, placentation axile; styles 5, distinct, persistent; stigmas 5, terminal on each style, usually capitate, punctate, or 2-lobed. Fruit usually a loculicidal capsule, often deeply 5-angled; seeds usually arillate and discharged from capsule by elastic separation of aril from testa; endosperm copious, oily, fleshy; embryo large, spatulate, straight, enveloped by endosperm.

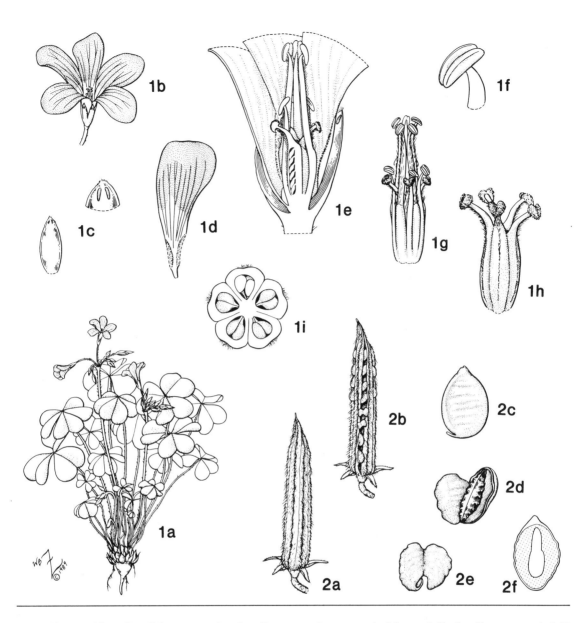

Figure 10. Oxalidaceae. 1, *Oxalis corymbosa:* **a,** habit, x 1/3; **b,** flower, x 1 1/2; **c,** sepal, x 3, with detail of apex showing dark calcium oxalate deposits, x 12; **d,** petal, x 3; **e,** longitudinal section of flower, x 7; **f,** anther, x 25; **g,** androecium and gynoecium, x 6; **h,** pistil, x 10; **i,** cross section of ovary, x 15. **2,** *Oxalis stricta:* **a,** capsule, x 2 1/2; **b,** dehiscing capsule, x 2 1/2; **c,** seed with aril, x 12; **d,** seed with dehiscing aril, x 12; **e,** aril after seed expelled, x 12; **f,** longitudinal section of seed, x 15.

Family characterization in Florida: herbaceous plants with bulbous or tuberous stems and acrid juice; palmately compound leaves in basal or apical clusters; leaflets with pulvini responsible for "sleep movements"; 5-merous, often heterostylous flowers; monadelphous stamens; 5, distinct, persistent styles; 5-angled loculicidal capsule; and seeds with arils separating elastically from testa. Acrid sap composed of oxalic acid (in the form of dissolved potassium oxalate; secreted as calcium oxalate often appearing on plant parts as white, red, or brown deposits). Anatomical feature: scattered secretory cavities in the mesophyll.

Genera/species: 7/890

Distribution: Primarily pantropical but also widespread in temperate regions.

Major genus: *Oxalis* (800 spp.)

Florida representatives: *Oxalis* (5-9 spp.)

Economic plants and products: Edible fruits from *Averrhoa* (carambola, bilimbi) and edible leaves and tubers from species of *Oxalis* (oca). Several weedy *Oxalis* spp. Ornamental plants: *Averrhoa*, *Biophytum* (life-plant), and *Oxalis* (Irish shamrock, lady's-sorrel, sheep-sorrel, wood-sorrel).

Some authors separate the woody members of the Oxalidaceae a segregate family. The large genus *Oxalis* is divided into sections and species according to characters of the leaflets (size, shape, and number), inflorescence (type), and corolla (color).

The wood-sorrels are easily identified in the field by their folded, "clover-like" leaves. As in the Fabaceae, the leaflets assume a "sleep position" (bend downwards) at night or in cold weather.

Heterostyly is common in the family (see Rubiaceae in *Part I* for explanation), especially in *Oxalis* species. Several *Oxalis* species are distylous and many are tristylous (Eiten, 1959, 1963; Denton, 1973; Robertson, 1975). The different style lengths of the tristylous flowers correspond to particular sets of filament lengths (in the biseriate androecium): long styles with medium + short filaments, medium length styles with long + short filaments, and short styles with long + medium filaments. Other differences may include pollen grain size, stigma and style morphology, and the pubescence of filaments and styles. The derivation and operation of the complex outcrossing systems in these *Oxalis* species has received much attention in the literature (Ornduff, 1972; Weller, 1976). Ideally, a legitimate cross may occur only between flowers with stamens and styles in a similar position (e.g., pollen from "long" stamen with long-styled stigma), although some tristylous *Oxalis* species are also self-compatible.

Oxalis flowers attract various insects (such as bees and butterflies) that visit for the nectar secreted at the filament bases into the (more or less) tubular flower. In many species, nectar guides (lines) also occur on the petals. Cleistogamous flowers, which resemble buds (see Violaceae), are prevalent in several species.

The loculicidal capsule of *Oxalis* appears septicidal because of the deep lobing between the carpels, which in many species, are incompletely fused. Each of the five carpels splits along the abaxial suture to expose the seeds. Any disturbance may then cause the explosive ejection of the seeds by means of their arils that rapidly split abaxially and turn inside out. This reaction is caused by the turgid cells along the inner surface of the aril.

References Cited

Denton, M.E. 1973. A monograph of *Oxalis*, section *Ionoxalis* (Oxalidaceae) in North America. Publ. Mus. Mich. State Univ. Biol. 4: 455-615.

Eiten, G. 1959. Taxonomy and regional variation of *Oxalis* section *Corniculatae*. 379 pp. Ph.D. Dissertation. Columbia Univ., New York, NY.

_____. 1963. Taxonomy and regional variation of *Oxalis* section *Corniculatae*. 1. Introduction, keys and synopsis of the species. Amer. Midl. Naturalist 69: 257-309.

Ornduff, R. 1972. The breakdown of trimorphic incompatibility in *Oxalis* section *Corniculatae*. Evolution 26: 52-65.

Robertson, K.R. 1975. The genera of Oxalidaceae in the southeastern United States. J. Arnold Arbor. 56: 223-239.

Weller, S.G. 1976. The genetic control of tristyly in *Oxalis* section *Ionoxalis*. Heredity 37: 387-393.

POLYGALACEAE (MILKWORT FAMILY)

Annual or perennial herbs, vines, or shrubs, often with taproots. Leaves simple, entire, usually alternate, sometimes reduced to scales, exstipulate or stipules represented by glands. Inflorescence indeterminate, capitate, spicate, racemose, or paniculate, or sometimes flower solitary, terminal or axillary. Flowers zygomorphic, perfect, hypogynous, each subtended by a bract and 2 bracteoles, sometimes with intrastaminal disc or nectariferous gland. Calyx of 5 sepals, distinct or the 2 lower (abaxial) connate, often with the 2 inner (lateral) sepals (wings) enlarged and petaloid, imbricate, persistent or caducous. Corolla of basically 5 petals but usually reduced to 3 (2 upper and 1 lower), variously connate and/or basally adnate to the androecium and forming a tube, with the abaxial (lower) petal (keel) often concave and crested with fringe, white, pink, yellow, or orange, imbricate. Androecium of usually 8 stamens; filaments connate into a tube-like sheath with an adaxial split above; anthers basifixed, often confluently 1-loculed, dehiscing by an apical or subapical pore or by a V-shaped slit, introrse. Gynoecium of 1 pistil, usually 2-carpellate; ovary superior, 2-loculed; ovules solitary in each locule, anatropous to hemitropous, pendulous, placentation axile; style 1, often apically bilobed with one lobe receptive and the other sterile and tufted with hairs; stigma capitate. Fruit usually a loculicidal capsule splitting into 2 1-seeded valves; seeds often with stiff and rigid hairs, with conspicuous aril-like outgrowth at the micropyle; endosperm usually copious, soft, fleshy, oily and proteinaceous; embryo straight.

Family characterization in Florida: herbaceous plants; brightly-colored papilionaceous-like flowers with modified perianths; calyx of 5 sepals with the 2 inner petaloid and wing-like; reduced corolla of 2 + 1 petals with the lower (keel) boat-shaped and fringed; 8 stamens with filaments connate into a split sheath; confluently 1-loculed anthers with poricidal dehiscence; bilobed style with a receptive lobe and a hairy, sterile lobe; and hairy seeds with micropylar outgrowths. Pollen grains distinctively polycolporate. Tissues commonly with calcium oxalate crystals. Anatomical features: closed ring of xylem in young stems and lysigenous secretory cavities or oil ducts (*Polygala*).

Genera/species: 15/800

Distribution: Cosmopolitan; absent from New Zealand and the Arctic (and with only a few introduced weeds in Polynesia).

Major genera: *Polygala* (500-600 spp.) and *Monnina* (150 spp.)

Florida representatives: *Polygala* (25 spp.); see Saulmon (1971).

Economic plants and products: Medicinal roots (due to saponins) from several *Polygala* spp., such as *P. senega* (snakeroot -- with seregin). Ornamental plants: *Polygala* (milkwort), *Monnina*, and *Securidaca*.

The Polygalaceae are a natural family, although several views have been proposed as to their systematic position (Miller, 1971). Tribes (and genera) are based mainly upon fruit morphology, with *Polygala* having capsular fruits.

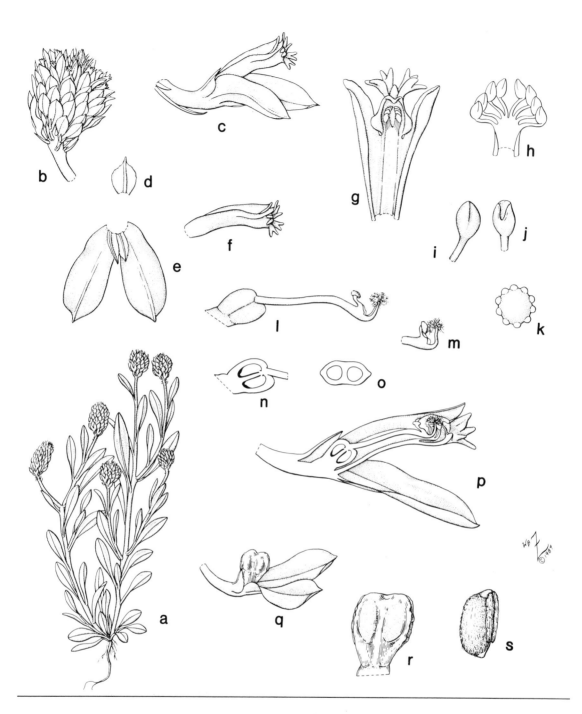

Figure 11. Polygalaceae. *Polygala lutea:* **a,** habit, x 1/2; **b,** inflorescence, x 2; **c,** flower, x 6; **d,** adaxial sepal, x 6; **e,** abaxial and lateral ("wing") sepals, x 6; **f,** corolla, x 6; **g,** expanded corolla and androecium, x 10; **h,** androecium, x 15; **i,** anther, x 36; **j,** dehisced anther, x 36; **k,** pollen grain (polycolporate), x 225; **l,** pistil showing orientation of receptive and sterile lobes of style before dehiscence of anthers, x 12; **m,** apex of style from older flower, x 12; **n,** longitudinal section of ovary, x 12; **o,** cross section of ovary, x 12; **p,** longitudinal section of flower, x 10; **q,** capsule with persistent calyx, x 4 1/2; **r,** capsule (calyx removed), x 9; **s,** seed, x 16.

A novice in the field could casually mistake a milkwort for an orchid or a legume. Actually, a *Polygala* flower does superficially resemble the papilionaceous flower of the Fabaceae, but the similar parts are not homologous. The conspicuous "wings" are the enlarged and petaloid lateral sepals. The corolla, reduced to three petals, is usually adnate to the staminal "tube" or sheath (slit on adaxial side). The lower median petal, often bearing a fringed crest, forms a "keel" closely surrounding the stigma and anthers (Holm, 1929).

The pollination of *Polygala* species is complex and little-studied. It is known that self-pollination may occur in our Florida species in which the sterile apical lobe of the stigma consists of a tuft of hairs that catch the pollen when the anthers open. As the flower develops, the sterile and the receptive lobes of the stigma may touch each other (see Fig. 11, 1 and m), resulting in the transfer of pollen. In several other species (not occurring in Florida) where insect pollination has been closely observed, the pollen accumulates in a trough-like or horizontal extension of the style, which lacks the hairy sterile tip of our species. Basically, insects (mainly bees) seek the nectar at the base of the flower and land upon the keel, which exposes the stigmas and anthers. The insect may then pick up pollen from the trough as it leaves and/or enters the flower.

The seeds of *Polygala* have 2- or 3-lobed aril-like outgrowths. The structure develops from the tissues of the outer integuments at the micropylar end of the seed and is, thus, not properly termed an aril (which develops from funiculus at the hilum) or a caruncle (which develops from integuments at the hilum). The seeds are distributed by ants, which utilize the outgrowths as a food source.

References Cited

Holm, T. 1929. Morphology of North American species of *Polygala*. Bot. Gaz. (Crawfordsville) 88: 167-185.

Miller, N.G. 1971. The genera of Polygalaceae in the southeastern United States. J. Arnold Arbor. 52: 267-284.

Saulmon, J.G. 1971. A revision of the Polygalaceae of the southeastern United States. 187 pp. Ph.D. Dissertation. West Virginia Univ., Morgantown, WV.

SALICACEAE (WILLOW FAMILY)

Trees, shrubs, or sometimes subshrubs. <u>Leaves</u> simple, often serrate, alternate, deciduous, stipulate (stipules often conspicuous and caducous). <u>Inflorescence</u> indeterminate, spicate, erect or pendulous (catkin), lateral. <u>Flowers</u> more or less actinomorphic, imperfect (plants dioecious), hypogynous, usually precocious, very reduced, each subtended by a fringed or hairy bract. <u>Perianth</u> absent or reduced to a cupular disc or 1 or 2 nectariferous glands. <u>Androecium</u> of 2 to numerous stamens; filaments distinct to basally connate; anthers basifixed, dehiscing longitudinally. <u>Gynoecium</u> of 1 pistil, 2- to 4-carpellate; ovary superior, 1-loculed, sometimes stalked; ovules generally numerous, anatropous, placentation parietal or basal; style 1 or absent; stigmas 2 to 4; often bifid or irregularly lobed. <u>Fruit</u> a 2- to 4-valved loculicidal capsule; seeds small, numerous, usually with an apical tuft of hairs (coma); endosperm absent or scanty and oily; embryo small, straight.

Family characterization in Florida: fast-growing and often clonal dioecious trees and shrubs; deciduous, stipulate leaves; reduced imperfect flowers subtended by fringed bracts and arranged in erect or pendent spicate inflorescences (catkins); absent or vestigial perianth represented by cup-like disc or gland(s); and comose seeds. Tissues with calcium oxalate crystals, a high tannin content, and certain glucosides (e.g., salicin and

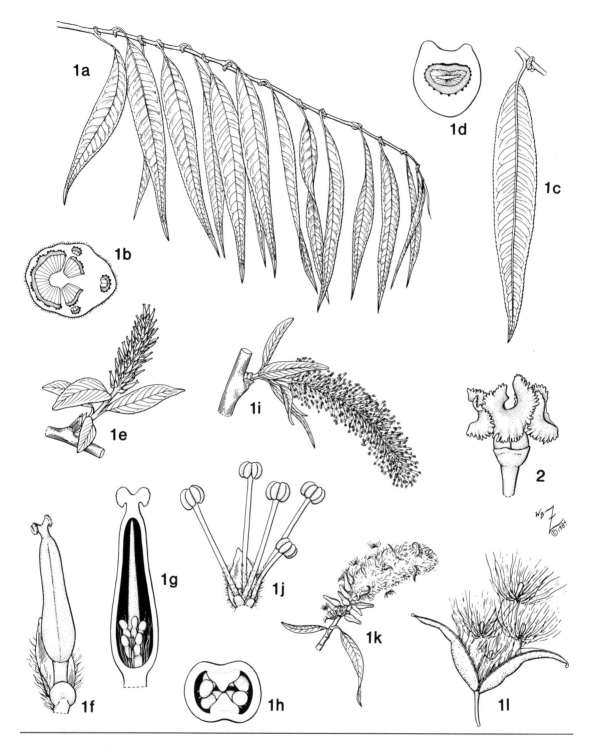

Figure 12. Salicaceae. 1, *Salix caroliniana:* **a,** sterile branch, x 1/3; **b,** cross section of node (three-trace trilacunar), x 6; **c,** leaf, x 1/2; **d,** cross section of petiole, x 18; **e,** carpellate inflorescence, x 1 1/4; **f,** carpellate flower and bract, x 10; **g,** longitudinal section of pistil, x 15; **h,** cross section of ovary, x 25; **i,** staminate inflorescence, x 1 1/4; **j,** staminate flower and bract, x 8; **k,** fruiting branch, x 3/4; **l,** capsule, x 4 1/2. **2,** *Populus deltoides:* carpellate flower, x 4 1/2.

populin). Anatomical features: periderm with superficial origin (from epidermis or the next lower layer), distal end of petioles with closed ring(s) of xylem and phloem, often unitegmic ovules (or second integument very thin), and trilacunar nodes.

Genera/species: 2/530

Distribution: Generally widespread with centers of diversity in north temperate and subarctic regions; absent in Australia and the Malay Archipelago. Common in moist habitats.

Major genus: *Salix* (about 500 spp.); species delimitations difficult in *Salix* (and also in *Populus*) due to extensive hybridization.

Florida representatives: *Salix* (5 spp.) and *Populus* (2 spp.)

Economic plants and products: Slender stems ("osiers") used in basketry from *Salix*. Lumber from species of *Populus* (wood pulp, boxes, matches) and *Salix* (boxes, matches). Medicinal bark from *Salix* (due to salicylic acid). Ornamental trees: *Salix* (willow) and *Populus* (cottonwood, aspen, poplar).

Since the Salicaceae are somewhat isolated taxonomically, their affinities to other groups are in some dispute (see Meeuse, 1975; Thorne, 1973, 1983; and Cronquist, 1981). In the past, the Salicaceae had been included in the "Amentiferae" (see Fagaceae and Betulaceae in *Part I*), an artificial assemblage of woody plants with reduced flowers arranged in aments (Hjelmquist, 1948; Stern, 1973). Actually, the apparent "simplicity" of the flowers is due to extreme reduction (Fisher, 1928).

The minute and imperfect flowers, each in the axil of a bract, are congested into erect or pendulous spicate inflorescences (aments, catkins) that are unisexual. The trees are dioecious. In *Populus*, the inflorescences are pendulous, and the flowers are entirely wind-pollinated. Although a disc- or cup-shaped gland (reduced perianth) occurs at the base of each flower, no nectar or scent is produced. The flowers of *Salix*, however, attract various insects (especially bees and moths) that visit to collect pollen and/or nectar. Both the staminate and carpellate flowers secrete abundant and sweetly scented nectar from glands at the flower bases. Since the flowers often appear before or at the same time as the leaves, the erect spikes of reduced flowers are somewhat conspicuous, especially the staminate inflorescences with the long filaments and bright yellow anthers. Although entomophily is prevalent in *Salix*, much pollen is also spread by the wind, which is probably also important in pollination.

References Cited

Cronquist, A. 1981. An integrated system of classification of flowering plants. Pp. 432-435. Columbia Univ. Press, New York, NY.

Fisher, M.J. 1928. The morphology and anatomy of flowers of Salicaceae. I. and II. Amer. J. Bot. 15: 307-326, 372-394.

Hjelmquist, H. 1948. Studies on the floral morphology and phylogeny of the Amentiferae. Bot. Not., Suppl., 2: 1-171.

Meeuse, A.D.J. 1975. Taxonomic relationships of Salicaceae and Flacourtiaceae. Acta Bot. Neerl. 24: 437-457.

Stern, W.L. 1973. Development of the amentiferous concept. Brittonia 25: 316-333.

Thorne, R.F. 1973. The "Amentiferae" or Hamamelidae as an artificial group: a summary statement. Brittonia 25: 395-405.

_____. 1983. Proposed new realignments in the angiosperms. Nord. J. Bot. 3: 85-117.

CUCURBITACEAE (PUMPKIN OR SQUASH FAMILY)

Annual or sometimes perennial vines, sometimes softly woody, climbing or prostrate and trailing, often coarse and scabrous, sometimes somewhat succulent with abundant watery sap, with swollen tuberous rootstock; stems often 5-angled in cross section. Leaves simple and palmately 5-lobed or sometimes palmately compound, alternate, palmately veined, with spirally coiled tendrils at petiole base, exstipulate. Inflorescence determinate, cymose or flowers solitary, axillary. Flowers actinomorphic, imperfect (plants monoecious or dioecious), epigynous, with cup- or tube-like expanded hypanthium, typically with variously developed nectary, ephemeral. Calyx synsepalous with 5 lobes, tubular, imbricate or open. Corolla sympetalous with 5 lobes, campanulate, rotate, to salverform, yellow or sometimes greenish, orange-yellow, or white, valvate or induplicate-valvate. Androecium of basically 5 stamens but often apparently 3 or sometimes 1 due to various modifications (cohesion, connation and/or convolution); filaments distinct or coherent to connate in 2 pairs (with the fifth one distinct) to completely connate (monadelphous); anthers 1- or seemingly 2-loculed, distinct or coherent to connate (in 2 pairs or completely connate), straight or often variously bent to convoluted, dehiscing longitudinally, extrorse; staminodes sometimes present in carpellate flowers. Gynoecium of 1 pistil, usually 3-carpellate; ovary inferior, typically 1-loculed; ovules numerous, anatropous, placentation fundamentally parietal (with 3 enlarged and bifid placentae almost completely filling the locule); style usually 1, columnar; stigmas usually 3, thick, each bilobed or bifid; rudimentary pistil often present in staminate flowers. Fruit usually a berry or a modified berry (pepo) with leathery or hard pericarp or sometimes a variously dehiscing capsule; seeds many, large, usually compressed, with many-layered seed coat; endosperm absent; embryo straight, oily, with large and flat cotyledons.

Family characterization in Florida: prostrate or climbing vines often with coarse and scabrous parts; 5-angled stems; palmately lobed or compound leaves with tendrils at the petiole base; yellow or white imperfect flowers; unusual stamens highly modified by displacement, reduction, fusion, and/or folding; inferior ovary with 3 greatly enlarged parietal placentae; and a berry (often a pepo) or fleshy capsule as the fruit type. Plants characteristically with cucurbitacins (tetracyclic triterpenoids) and a high alkaloid content. Anatomical features: trichomes often with calcified walls and with cystoliths (at the base and in nearby cells) and bicollateral vascular bundles in the stem (often arranged in 2 concentric rings) and petiole.

Genera/species: 110/640

Distribution: Pantropical and subtropical; a few representatives in temperate to cooler climates.

Major genera: *Gurania* (75 spp.), *Cayaponia* (45 spp.), and *Momordica* (45 spp.)

Florida representatives: 10 genera/14 spp.; largest genera: *Cucumis* (3 spp.), *Cucurbita* (2 spp.), and *Momordica* (2 spp.)

Economic plants and products: Food plants: *Citrullus* (watermelon), *Cucurbita* (gourds, pumpkins, squashes, vegetable spaghetti), *Cucumis* (cantaloupe, cucumber, gherkin, honeydew, muskmelon), *Lagenaria* (calabash), *Momordica* (balsam-apple, bitter melon), and *Sechium* (chayote). Loofa sponges (dried vascular system of fruit) from *Luffa*. Ornamental plants (species of 22 genera), including: *Benincasa* (Chinese-watermelon), *Coccinea* (ivy-gourd), *Ecballium* (squirting-cucumber), *Lagenaria*, *Luffa* (vegetable-sponge), *Sicana* (cassabanana), *Sicyos* (bur-cucumber), and *Trichosanthes* (snake-gourd).

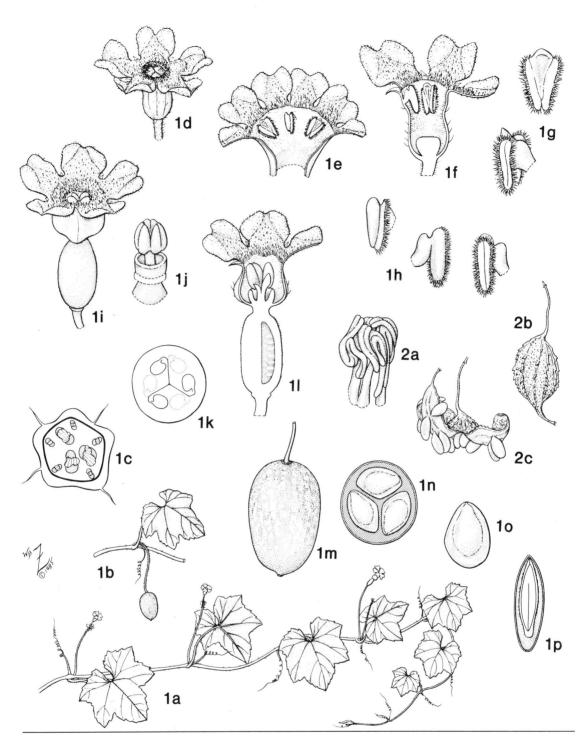

Figure 13. Cucurbitaceae. 1, *Melothria pendula:* **a,** habit with flowers, x 1/2; **b,** habit with mature fruit, x 1/2; **c,** cross section of stem showing bicollateral bundles, x 18; **d,** staminate flower, x 5; **e,** expanded corolla and androecium of staminate flower, x 6; **f,** longitudinal section of staminate flower, x 7 1/2; **g,** two views of a pair of connate stamens, x 15; **h,** three views of the single stamen, x 15; **i,** carpellate flower, x 5; **j,** detail of apical portion of pistil showing style, stigma, and nectary, x 7 1/2; **k,** cross section of ovary, x 10; **l,** longitudinal section of carpellate flower, x 6; **m,** berry, x 2; **n,** cross section of berry, x 2; **o,** seed with mucilaginous outer seed coat, x 3; **p,** longitudinal section of seed, x 4 1/2. **2,** *Momordica charantia:* **a,** androecium, x 6; **b,** fruit, x 2/3; **c,** dehisced fruit, x 2/3.

The Cucurbitaceae compose a distinctive and specialized family in habit, floral structure, and biochemistry. There are several differing opinions as to their phyletic position (see Lawrence, 1951 and Cronquist, 1981).

The unusual androecium of the cucurbits basically consists of five stamens that vary in the connation of the filaments and anthers (Chakravarty, 1958). Key characters of genera are often based upon the morphology of the androecium (Jeffrey, 1962, 1967, 1980). The most common situation (e.g., *Melothria*) is an androecium of apparently "three" stamens, two with four pollen sacs and one with two pollen sacs. Each two-loculed stamen (with four pollen sacs) is actually a compound stamen formed by the fusion of two adjacent stamens. Other variations include contorted "capitate" anthers (as in *Momordica*) and/or monadelphous stamens as in species of *Cucurbita* or *Sicyos*.

The flowers attract various insects (often bees) that visit to collect the nectar or sometimes also the copious pollen. Cross-pollination is favored by the monoecious or dioecious nature of the plants, and in monoecious species, the staminate flowers often appear (on the younger parts of the stem) before the carpellate flowers.

The Cucurbitaceae are also characterized by distinctive fruits and seeds. The fruit is usually a berry, although other types occur (such as fleshy capsules). When modified with a firm-walled epicarp (rind), the berry is termed a "pepo." The seed has a several-layered seed coat, and the outer layers often swell in water (Corner, 1976). The outer-most covering is derived from the carpel wall (endocarp), and the inner layers, from the outer integument.

References Cited

Chakravarty, H.L. 1958. Morphology of the staminate flowers in the Cucurbitaceae with special reference to the evolution of the stamens. Lloydia 21: 49-87.

Corner, E.J.H. 1976. The seeds of dicotyledons. Pp. 112-114. Cambridge Univ. Press, Cambridge, England.

Cronquist, A. 1981. An integrated system of classification of flowering plants. Pp. 422-425. Columbia Univ. Press, New York, NY.

Jeffrey, C. 1962. Notes on Cucurbitaceae, including a proposed new classification of the family. Kew Bull. 15: 337-371.

_____. 1967. On the classification of the Cucurbitaceae. Kew Bull. 20: 417-426.

_____. 1980. A review of the Cucurbitaceae. J. Linn. Soc., Bot. 81: 233-247.

Lawrence, G.H.M. 1951. Taxonomy of vascular plants. Pp. 718-720. Macmillan Publishing Co., Inc., New York, NY.

ULMACEAE (ELM FAMILY)

Trees and shrubs, with watery to mucilaginous sap. <u>Leaves</u> simple, entire to variously serrate, usually with obliquely-based blades, alternate, commonly distichous, stipulate (stipules caducous). <u>Inflorescence</u> determinate, cymose and of congested fascicles, or flower solitary, axillary. <u>Flowers</u> actinomorphic to slightly zygomorphic, perfect or imperfect (then plants usually monoecious), hypogynous, small and inconspicuous. <u>Perianth</u> of generally 4 to 8 tepals, distinct to connate, campanulate, foliaceous, imbricate, persistent. <u>Androecium</u> with as many stamens as tepals and opposite them, free or adnate to the perianth base; filaments distinct, erect in bud; anthers basifixed, dehiscing longitudinally, extrorse or introrse. <u>Gynoecium</u> of 1 pistil, 2-carpellate; ovary superior, usually 1-loculed, sessile to stalked; ovule solitary, anatropous or amphitropous, pendulous, placentation apical; styles 2, linear, stigmas decurrent and along upper inner surface

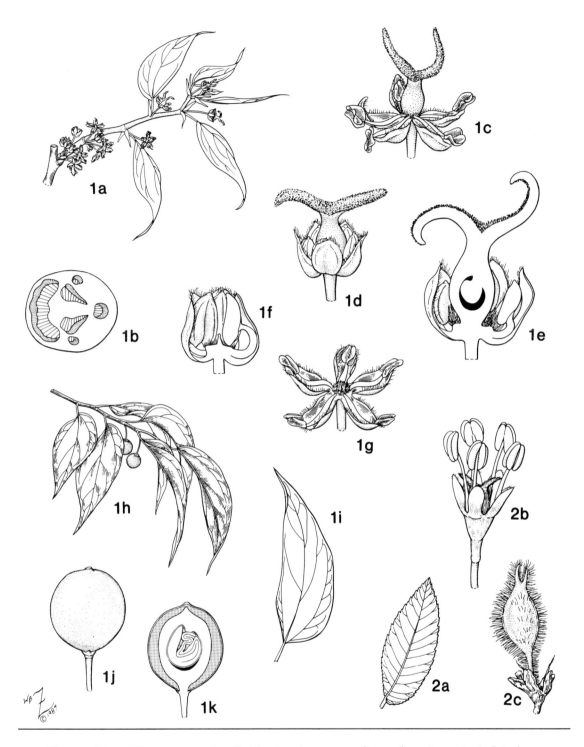

Figure 14. Ulmaceae. 1, *Celtis laevigata:* **a,** flowering branch (with immature leaves), x 1; **b,** cross section of node (three-trace trilacunar), x 8; **c,** perfect flower, x 6; **d,** carpellate flower, x 6; **e,** longitudinal section of carpellate flower, x 10; **f,** longitudinal section of bud of staminate flower, x 10; **g,** staminate flower, x 6; **h,** fruiting branch, x 1/2; **i,** mature leaf, x 1/3; **j,** drupe, x 3; **k,** longitudinal section of drupe, x 3. **2,** *Ulmus alata:* **a,** leaf, x 3/4; **b,** flower, x 7; **c,** samara, x 4 1/2.

of styles; rudimentary pistil often present in staminate flowers. Fruit a nutlet, samara, or drupe; endosperm usually absent or scanty; embryo straight or curved.

Family characterization in Florida: trees or shrubs with watery to slightly mucilaginous sap; stipulate leaves with oblique bases; reduced flowers with uniseriate perianths; stamens with erect filaments in bud; a samara or drupe as the fruit type; and seeds with little or no endosperm. Anatomical features: mucilage cells and/or canals in the tissues, often calcification or silicification of certain cell walls (epidermal hairs of stem and leaves), and trilacunar nodes.

Genera/species: 15/200

Distribution: Primarily throughout temperate and tropical regions of the Northern Hemisphere.

Major genera: *Celtis* (80 spp.), *Ulmus* (25-30 spp.), and *Trema* (30 spp.)

Florida representatives: *Celtis* (4 spp.), *Ulmus* (4 spp.), *Trema* (2 spp.), and *Planera* (1 sp.)

Economic plants and products: Timber from several, such as species of *Ulmus* (elm). Medicinal bark from *Ulmus rubra* (slippery elm -- with high mucilage content). Edible fruit from *Celtis* (hackberries, sugarberries). Ornamental trees and shrubs (species of 8 genera), including: *Ulmus, Celtis, Planera* (water-elm, planer-tree), and *Zelkova.*

The Ulmaceae are divided into two very distinctive tribes or subfamilies (Ulmoideae and Celtidoideae) on the basis of numerous morphological, chemical, and anatomical features (Elias, 1970; Sweitzer, 1971; Giannasi, 1978; Cronquist, 1981). For example, leaf, fruit, and seed characters, which are often cited, differ in the two groups as follows:

Ulmoideae (*Ulmus, Planera*) - pinnately veined leaves with secondary veins running to the teeth, dry (often winged) fruit, flat seeds, straight embryo with flat cotyledons, no endosperm;

Celtidoideae (*Celtis, Trema*) - venation usually with three main veins diverging from the base and secondary veins forming a series of arches, baccate fruit (drupes), round seeds with folded or rolled cotyledons, some endosperm usually present.

In addition, the two groups are characterized by different (but overlapping) distributions, with the Ulmoideae mainly north temperate, and the Celtidoideae, typically tropical to subtropical in range.

The reduced and relatively inconspicuous flowers of the Ulmaceae are anemophilous and bloom early in the season (Berg, 1977). Most species have imperfect flowers and are monoecious. A notable exception is *Ulmus* (flowers usually perfect). Often the carpellate flowers on a tree will develop before the staminate flowers (and also before the perfect flowers, if also present). The mechanisms for preventing possible self-pollination in *Ulmus* flowers are not well understood.

References Cited

Berg, C.C. 1977. Urticales, their differentiation and systematic position. Pl. Syst. Evol., Suppl. 1: 349-374.

Cronquist, A. 1981. An integrated system of classification of flowering plants. Pp. 189-193. Columbia Univ. Press, New York, NY.

Elias, T.S. 1970. The genera of Ulmaceae in the southeastern United States. J. Arnold Arbor. 51: 18-40.

Giannasi, D.E. 1978. Generic relationships in the Ulmaceae based on flavonoid chemistry. Taxon 27: 331-344.

Sweitzer, E.M. 1971. Comparative anatomy of Ulmaceae. J. Arnold Arbor. 52: 523-585.

MORACEAE (MULBERRY FAMILY)

Usually trees or shrubs, with milky sap, often with glandular hairs. <u>Leaves</u> simple, entire, serrate, or lobed, usually alternate, often with 3 to 5 palmate veins at base, deciduous or persistent, stipulate (stipules caducous, small or cap-like, and leaving a circular scar). <u>Inflorescence</u> determinate, basically cymose and often appearing racemose, spicate (pendulous), umbellate, or capitate/globose, or flowers along inner surface of involuted or invaginated hollow receptacle, axillary. <u>Flowers</u> actinomorphic, imperfect (then plants monoecious or dioecious), hypogynous to epigynous, minute. <u>Perianth</u> of usually 4 tepals or sometimes reduced or absent, biseriate, distinct to connate, imbricate or valvate, often persistent and accrescent. <u>Androecium</u> of usually 4 stamens, opposite the tepals; filaments distinct, erect or inflexed in bud; anthers versatile, dehiscing longitudinally. <u>Gynoecium</u> of 1 pistil, 2-carpellate with usually 1 carpel aborting; ovary superior to inferior, usually 1-loculed; ovule solitary, usually anatropous, pendulous, placentation apical or subapical; styles usually 2, filiform; stigmas 2; rudimentary pistil sometimes present in staminate flowers. <u>Fruit</u> a drupe (sometimes dehiscent) or sometimes an achene, usually aggregated into a multiple fruit (syncarp) arising from the union of fruits of different flowers, their perianths, and common receptacle; endosperm fleshy and oily or absent; embryo curved, often with unequal cotyledons.

Family characterization in Florida: trees or shrubs with usually milky sap; reduced, imperfect flowers in highly modified cymose inflorescences; 4-merous perianth and androecium; and individual drupes aggregated into a multiple fruit (syncarp). Anatomical features: laticifers (containing various substances) in stems and leaves; often calcification or silicification of certain cell walls (epidermal hairs and tissues of leaves); cystoliths in leaf epidermal cells; and trilacunar to multilacunar nodes.

Genera/species: Moroideae: 53/1,400; Urticaceae *s.l.*: 98/2,400 (see Urticaceae in *Part I* and Thorne, 1983).

Distribution: Widespread in tropical and subtropical regions with a few representatives in temperate areas.

Major genera: *Ficus* (800 spp.), *Dorstenia* (170 spp.), and *Cecropia* (100 spp.)

Florida representatives: *Ficus* (5 spp.), *Morus* (2 spp.), *Brosimum* (1 sp.), *Broussonetia* (1 sp.), *Dorstenia* (1 sp.), *Fatoua* (1 sp.), and *Maclura* (1 sp.)

Economic plants and products: Edible fruits from: *Artocarpus* (breadfruit, jackfruit), *Ficus* (figs), and *Morus* (mulberries). Timber from *Chlorophora* (fustic, iroko-wood) and *Maclura* (osage-orange). Rubber from latex of several, such as species of *Castilla* and *Ficus*. Ornamental plants (species of 13 genera), including: *Broussonetia* (paper-mulberry), *Cecropia*, *Chlorophora*, *Cudrania*, *Dorstenia* (pickaback-plant), *Ficus* (various figs, India rubber-plant, banyan), and *Maclura*.

38

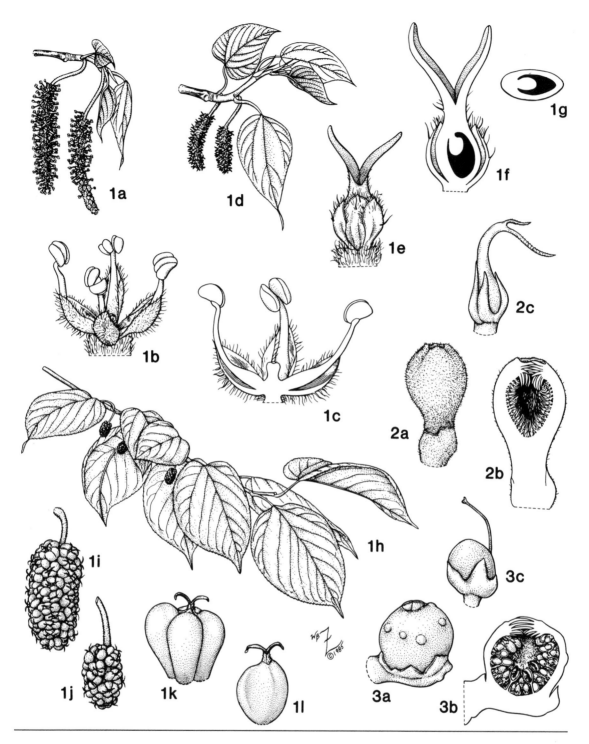

Figure 15. Moraceae. 1, *Morus rubra:* **a,** branch with staminate inflorescences, x 3/4; **b,** staminate flower, x 8; **c,** longitudinal section of staminate flower, x 9; **d,** branch with carpellate inflorescences, x 3/4; **e,** carpellate flower, x 12; **f,** longitudinal section of carpellate flower, x 20; **g,** cross section of ovary, x 25; **h,** fruiting branch, x 1/3; **i,** syncarp (from cultivated specimen), x 1 1/2; **j,** syncarp (from wild specimen), x 1 1/2; **k,** drupe with accrescent perianth, x 6; **l,** drupe (perianth removed), x 6. **2,** *Ficus carica:* **a,** inflorescence (syconium), x 3; **b,** longitudinal section of syconium, x 4; **c,** carpellate flower, x 25. **3,** *Ficus aurea:* **a,** syncarp ("fig"), x 3; **b,** longitudinal section of syncarp (many wasps have been removed), x 4; **c,** drupe with persistent perianth, x 20.

The delimitation of the Moraceae varies considerably in the literature (see Tippo, 1938; Berg, 1977; and Cronquist, 1981). Basically, the Moraceae *s.s.* have been united with or segregated from the closely related Urticaceae, Cecropioideae (Conocephaloideae, Cecropiaceae), and Cannabaceae. For example, Thorne (1983) includes the Moraceae as a subfamily (Moroideae) of the Urticaceae *s.l.* (see *Part I*), along with the Cecropioideae. The Urticoideae and Moroideae are connected by transitional genera (Heywood, 1978) and are basically distinguished by the number of styles (one or two), placentation (basal or apical), and type of sap (clear vs. milky). The Cannabaceae and the Cecropioideae have been considered as segregate families or as subfamilies within either the Moraceae or the Urticaceae (Gangadhara and Inamdar, 1977; Berg, 1978).

The inflorescence type is important in classification within the Moraceae. For example, *Morus* is characterized by staminate and carpellate catkins. The carpellate flowers of *Maclura* and *Broussonetia* are congested into globose heads. In *Ficus*, an involuted receptacle becomes a hollow and fleshy structure (syconium) that bears the flowers along the interior surface.

The flowers of the Moraceae are usually anemophilous (as in *Morus*), although the specialized and very complex entomophilous pollination of *Ficus* by gall-wasps has received much attention in the literature (Proctor and Yeo, 1972; Faegri and Pijl, 1979). Basically, the flowers (enclosed in the hollow, fleshy axis) are pollinated when the wasp enters to lay eggs in the ovaries of the carpellate flowers. Newly emergent wasps carry pollen from that inflorescence to new syconia. The development of the carpellate flowers before the staminate flowers within the same inflorescence encourages cross-pollination.

A characteristic multiple fruit of the Moraceae is usually composed of drupes (from adjacent flowers), accrescent perianths, and the fleshy common receptacle. The individual drupes and subtending perianths may be easily distinguished in a mulberry (*Morus*), which superficially resembles a blackberry (aggregate fruit of species of *Rubus*). In a breadfruit (*Artocarpus*), the drupes and axis are well-united into one massive structure. A fleshy receptacle with drupes inside forms a fig (fruit of *Ficus*).

References Cited

Berg, C.C. 1977. Urticales, their differentiation and systematic position. Pl. Syst. Evol., Suppl. 1: 349-374.

_____. 1978. Cecropiaceae, a new family of the Urticales. Taxon 27: 39-44.

Cronquist, A. 1981. An integrated system of classification of flowering plants. Pp. 183-201. Columbia Univ. Press, New York, NY.

Faegri, K. and L. van der Pijl. 1979. The principles of pollination ecology. Pp. 176-178. Pergamon Press, Oxford, England.

Gangadhara, M. and J.A. Inamdar. 1977. Trichomes and stomata, and their taxonomic significance in the Urticales. Pl. Syst. Evol. 127: 121-137.

Heywood, V.H. (ed.) 1978. Flowering plants of the world. Pp. 96-97. Mayflower Books, New York, NY.

Proctor, M. and P. Yeo. 1972. The pollination of flowers. Pp. 312-316. Taplinger Publishing Co., New York, NY.

Thorne, R.F. 1983. Proposed new realignments in the angiosperms. Nord. J. Bot. 3: 85-117.

Tippo, O. 1938. Comparative anatomy of the Moraceae and their presumed allies. Bot. Gaz. (Crawfordsville) 100: 1-99.

ANACARDIACEAE (SUMAC OR CASHEW FAMILY)

Trees, shrubs, or occasionally woody vines, with resinous bark, branches, leaves, flowers, and fruits, sometimes with milky sap. Leaves trifoliolate, pinnately compound, or sometimes simple, entire to serrate, alternate, deciduous or persistent, usually exstipulate. Inflorescence determinate, basically cymose and appearing paniculate, terminal or axillary. Flowers actinomorphic, perfect or more often imperfect (then plants dioecious, polygamous, or polygamodioecious), hypogynous or rarely perigynous or epigynous, small, with annular intrastaminal nectariferous disc. Calyx of typically 5 sepals, usually basally connate, usually imbricate. Corolla of typically 5 petals, distinct or basally connate, white, green, or yellow, usually imbricate. Androecium of usually 5 or 10 (biseriate) stamens, arising upon or outside the disc; filaments distinct or rarely basally connate; anthers dorsifixed, versatile, dehiscing longitudinally, introrse; staminodes often present in carpellate flowers. Gynoecium of 1 pistil, usually 3-carpellate (2 carpels usually aborting); ovary usually superior, usually 1-loculed; ovule solitary, anatropous, often with thickened funiculus, placentation basal, parietal, or apical; style(s) 1 to 3, often widely separated; stigma(s) 1 to 3; rudimentary pistil often present in staminate flowers. Fruit usually a drupe with resinous mesocarp; endosperm scanty or absent; embryo oily, usually curved.

Family characterization in Florida: trees and shrubs with resinous parts; trifoliolate or pinnately compound leaves; often imperfect flowers with intrastaminal nectariferous disc and 5-merous perianth and androecium; 3-carpellate and 1-loculed ovary with a solitary ovule; and drupe with resinous mesocarp. Tissues characteristically with calcium oxalate crystals and a high tannin content. Anatomical features: well-developed schizogenous or lysigynous resin canals (exuding material that turns black on drying and that often contains irritant substances) or sometimes latex channels in the bark, leaves, flower, and/or fruits; and trilacunar nodes.

Genera/species: 70/600

Distribution: Mainly pantropical with a few representatives extending into temperate areas of Eurasia and North America.

Major genera: *Rhus* (100-250 spp.) and *Terminthia* (70 spp.)

Florida representatives: 6 genera/10 spp.; largest genera: *Toxicodendron* (3 spp.) and *Rhus* (3 spp.)

Economic plants and products: Edible seeds or fruits from: *Anacardium* (cashew), *Harpephyllum* (kaffir-plum), *Mangifera* (mango), *Pistacia* (pistachio), and *Spondias* (mombin, Jamaica-plum, hog-plum). Timber from *Schinopsis* (quebracho) and *Astronium* (kingwood, zebrawood). Resins, oils, and lacquers from several, such as species of *Pistacia* (mastic-tree) and *Toxicodendron* (varnish-tree). Tannic acid from: *Cotinus*, *Pistacia*, *Rhus*, and *Schinopsis*. Many toxic plants (due to irritant phenolic compounds; see discussion below): *Toxicodendron* (poison-ivy, -oak, and -sumac) and *Metopium* (poisonwood). Ornamental plants (species of 15 genera), including: *Cotinus* (smoke-tree), *Mangifera*, *Rhus* (various sumacs), and *Schinus* (Brazilian pepper-tree; but see Morton, 1979).

The Anacardiaceae are divided into five tribes based mainly upon the number and degree of fusion of the carpels. A notable taxonomic problem within the family is the delimitation of the heterogeneous *Rhus-Toxicodendron* complex (100 to 250 spp.), which has a confusing history (see Brizicky, 1962, 1963 and Gillis, 1971).

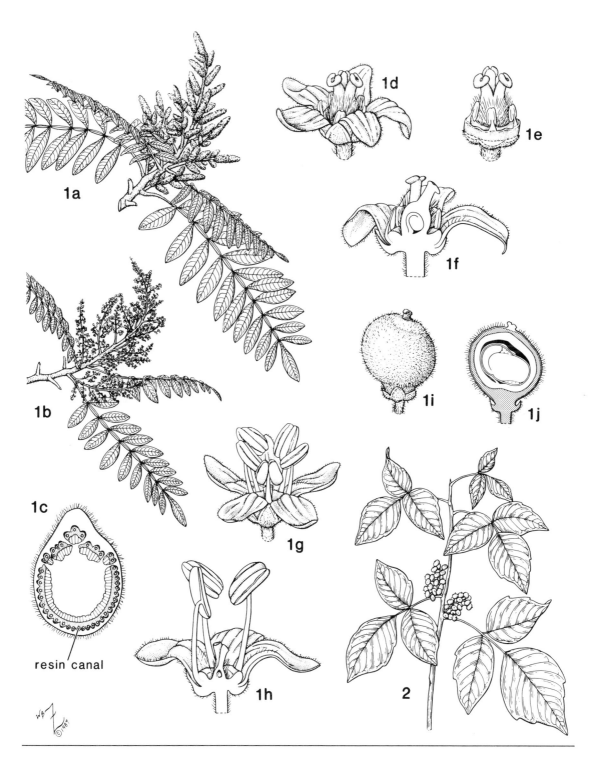

Figure 16. Anacardiaceae. 1, *Rhus copallina:* **a,** carpellate flowering branch, x 1/3; **b,** staminate flowering branch, x 1/3; **c,** cross section of node (three-trace trilacunar), x 3; **d,** carpellate flower, x 12; **e,** androecium and gynoecium of carpellate flower, x 14; **f,** longitudinal section of carpellate flower, x 15; **g,** staminate flower, x 9; **h,** longitudinal section of staminate flower, x 12; **i,** drupe, x 5; **j,** longitudinal section of drupe, x 5. **2,** *Toxicodendron radicans* (*R. radicans*): fruiting branch, x 1/3.

The characteristic resin canals of the family often produce exudate with poisonous compounds (urushiols) that may cause severe skin irritation. Direct or indirect contact as well as volatile emanation (from burning the plants) may cause an allergic reaction in sensitive individuals. The irritant substances may be distributed throughout various parts of the plant or concentrated in particular organs. For example, all parts of several *Toxicodendron* species (poison-ivy, -oak, and -sumac) and *Metopium* (all species) are allergenic. The lacquer produced from *T. vernicifluum* and the black ink from *Semecarpus anacardium* (marking-nut-tree) may cause allergic reactions. Although the flesh of the mango fruit (*Mangifera*) is edible, the skin may be an irritant to some people. The potentially irritating oil of *Anacardium occidentale* (cashew) is rendered harmless by heat.

The flowers of the Anacardiaceae are usually entomophilous. Various insects are attracted to the small flowers (in large inflorescences) with exposed nectar secreted by the disc. The dioecious and polygamodioecious nature of the plants promotes cross-pollination.

References Cited

Brizicky, G.K. 1962. The genera of Anacardiaceae in the southeastern United States. J. Arnold Arbor. 43: 359-375.

_____. 1963. Taxonomic and nomenclatural notes on the genus *Rhus* (Anacardiaceae). J. Arnold Arbor. 44: 60-80.

Gillis, W.T. 1971. The systematics and ecology of poison-ivy and the poison-oaks (*Toxicodendron*, Anacardiaceae). Rhodora 73: 72-159, 161-237, 370-443, 465-540.

Morton, J.F. 1979. Brazilian pepper: its impact on people, animals, and the environment. Econ. Bot. 32: 353-359.

JUGLANDACEAE (WALNUT FAMILY)

Usually trees, aromatic and resinous. Leaves pinnately compound, usually alternate, deciduous, glandular-dotted beneath (lepidote) and aromatic, exstipulate. Inflorescence basically determinate, generally spicate (erect or pendulous), with spikes sometimes grouped into panicles, few-flowered (carpellate inflorescences) to many-flowered (staminate inflorescences), terminal or axillary. Flowers actinomorphic, imperfect (plants usually monoecious), epigynous, small and inconspicuous, typically associated with bracts. Perianth of typically 4 tepals, sometimes modified into a disc or absent, distinct, minute, scale-like, free or adnate to subtending bracts, persistent. Androecium of 3 to many stamens; filaments distinct, short; anthers basifixed, dehiscing longitudinally. Gynoecium of 1 pistil, typically 2-carpellate; ovary inferior, 1-loculed above and usually 2-loculed at base (or apparently 4- to 8-loculed due to "false" partitions), adnate (at least at base) to involucre; ovule solitary, orthotropous, erect, appearing basal (especially in young flowers) but actually at the apex of the incomplete septum; styles 2, distinct or often basally connate, short, fleshy; stigmas 2, along inner sides of style branches, often plumose or papillate; rudimentary pistil sometimes present in staminate flowers. Fruit a drupe-like nut with dehiscent or indehiscent, leathery to fibrous husk (fused involucre and perianth); seed solitary, large; endosperm absent; embryo oily, often massive, often with variously lobed and corrugated cotyledons.

Family characterization in Florida: aromatic and resinous trees with pinnately compound glandular-dotted leaves; usually spicate, erect (carpellate) or pendulous (staminate) inflorescences; reduced imperfect flowers with uniseriate or absent perianth and associated bracts; inferior ovary with solitary "basal" ovule; drupe-like nut with leathery to fibrous husk; and large non-endospermous seeds with lobed and corrugated cotyledons.

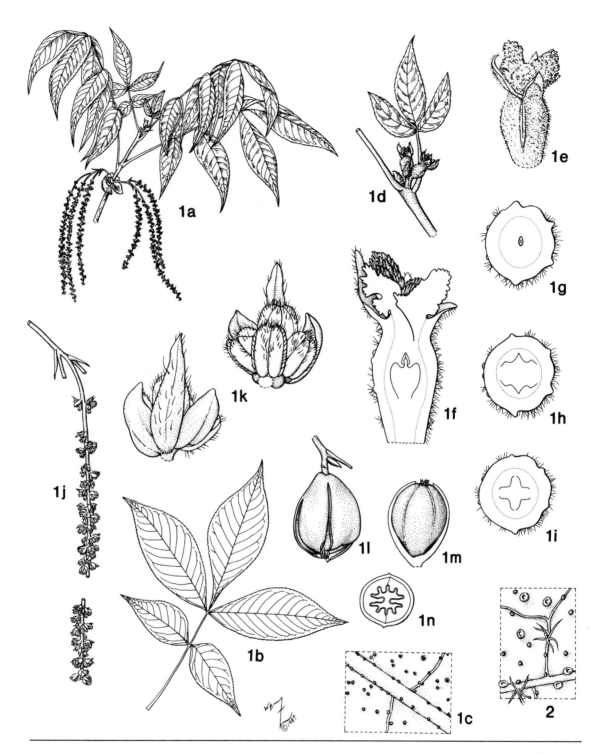

Figure 17. Juglandaceae. 1, *Carya glabra:* **a,** flowering branch, x 1/3; **b,** mature leaf, x 1/3; **c,** detail of abaxial surface of leaf showing peltate glands, x 25; **d,** carpellate inflorescence, x 1 1/3; **e,** carpellate flower, x 3; **f,** longitudinal section of carpellate flower, x 5; **g,** cross section of ovary (1-loculed) at level of ovule, x 5; **h,** cross section of ovary (2-loculed) near middle, x 5; **i,** cross section of ovary (4-loculed) near base, x 5; **j,** staminate inflorescence, x 1 1/3; **k,** two views of staminate flower, x 10; **l,** fruit (nut subtended by dehiscing husk), x 1/2; **m,** fruit (half of husk removed), x 1/2; **n,** cross section of nut, x 1/2. **2,** *Carya tomentosa:* detail of abaxial surface of leaf showing peltate glands (and stellate hairs), x 25.

Tissues commonly with tannins and calcium oxalate crystals. Anatomical features: peltate glands (lepidote) on leaves that secrete resinous substances; trilacunar nodes; and unitegmic ovules.

Genera/species: 7/59

Distribution: Primarily in temperate to subtropical regions of the Northern Hemisphere; also extending into montane areas of tropical America and tropical Asia. Most abundant in eastern Asia and eastern North America.

Major genera: *Carya* (25 spp.) and *Juglans* (15 spp.)

Florida representatives: *Carya* (8 spp.) and *Juglans* (1 sp.)

Economic plants and products: Edible seeds from *Carya* (pecans, hickory nuts) and *Juglans* (walnuts, butternuts). Valuable wood from: *Carya*, *Englehardtia*, and *Juglans*. Ornamental plants (species of 5 genera), including: *Carya*, *Juglans*, and *Pterocarya* (wingnut).

The phylogenetic relationships of the Juglandaceae to other families are in some dispute (see Elias, 1972; Cronquist, 1981; and Thorne, 1983). The family has been generally divided into two subfamilies (Manning, 1978) based upon the fruit morphology (drupe-like vs. samara-like). The genera *Juglans* and *Carya* are considered advanced within the family (e.g., studies by Manning, 1938, 1940, 1948), and are usually keyed to species on the basis of leaf characters.

The imperfect flowers of *Juglans* and *Carya*, each often subtended by three bracts, form unisexual inflorescences. The trees are usually monoecious. The flowers are anemophilous and open in the late spring after the leaves unfold. As in the Fagaceae, the ovules are not developed at the time of pollination.

The staminate aments, sometimes grouped into panicles, occur on the branches of the previous year. The staminate flowers vary in the number of stamens and the development of the perianth (e.g., no perianth in the flowers of *Carya*). The carpellate flowers form a few-flowered and erect inflorescence that is terminal on the current year's leafy shoot (Abbe, 1974). The bracts, which are variously united with the ovary (and perianth, when present), form a cupulate involucre that develops into a husk in the mature fruit (Manning, 1940; Hjelmquist, 1948).

The drupe-like fruits are dispersed by animals (Stone, 1973). In the fruit of *Carya*, the husk usually splits regularly into four valves, revealing a smooth nut. The husk of the fruit of *Juglans*, containing a sculptured nut, is more fleshy and dehisces irregularly by decaying. It is of interest that the split of the walnut "shell" corresponds to the midline of the two carpels, and not to the suture of the carpels.

References Cited

Abbe, E.C. 1974. Flowers and inflorescences of the 'Amentiferae'. Bot. Rev. 40: 159-261.

Cronquist, A. 1981. An integrated system of classification of flowering plants. Pp. 204-213. Columbia Univ. Press, New York, NY.

Elias, T.S. 1972. The genera of Juglandaceae in the southeastern United States. J. Arnold Arbor. 53: 26-51.

Hjelmquist, H. 1948. Studies on the floral morphology and phylogeny of the Amentiferae. Bot. Not., Suppl. 2: 1-171.

Manning, W.E. 1938. The morphology of the flower of the Juglandaceae. I. The inflorescence. Amer. J. Bot. 25: 407-419.

_____. 1940. Ibid. II. The pistillate flowers and fruit. Amer. J. Bot. 27: 839-852.

_____. 1948. Ibid. III. The staminate flowers. Amer. J. Bot. 35: 606-621.

_____. 1978. The classification within the Juglandaceae. Ann. Missouri Bot. Gard. 65: 1058-1087.

Stone, D.E. 1973. Patterns in the evolution of amentiferous fruits. Brittonia 25: 371-384.

Thorne, R.F. 1983. Proposed new realignments in the angiosperms. Nord. J. Bot. 3: 85-117.

ACERACEAE (MAPLE FAMILY)

Trees or sometimes shrubs, sometimes with milky sap. Leaves usually simple and palmately lobed with palmate venation or occasionally trifoliolate or pinnately compound, entire to deeply toothed, opposite, deciduous, exstipulate. Inflorescence basically determinate, corymbose, umbellate (in fascicles), racemose, or paniculate, usually terminal. Flowers actinomorphic, perfect or usually imperfect (then plants monoecious, dioecious, or polygamodioecious), hypogynous or sometimes perigynous, small, often with extrastaminal or intrastaminal nectariferous disc. Calyx of 4 or usually 5 sepals, distinct or basally connate, imbricate. Corolla of 4 or usually 5 petals, occasionally absent, distinct, often similar to calyx, greenish yellow to yellow or red, imbricate. Androecium of typically 8 or sometimes 4, 5, or 10 stamens, short and generally abortive in carpellate flowers; filaments distinct, arising on or inside the edge of disc; anthers basifixed, dehiscing longitudinally, introrse. Gynoecium of 1 pistil, 2-carpellate; ovary superior, 2-loculed, usually compressed perpendicular to the septum; ovules 2 in each locule, orthotropous, anatropous, campylotropous, or amphitropous, placentation axile; styles 2, distinct or basally connate, often recurved; stigmas 2, along inner surface of styles; rudimentary pistil often present in staminate flowers. Fruit a winged schizocarp ("double samara" or "paired samara") composed of 2 1-seeded and -winged segments that separate from a persistent carpophore or septum; endosperm absent; embryo conduplicate.

Family characterization in Florida: trees or shrubs with opposite, usually simple leaves with palmate lobing and venation; small imperfect flowers with a well-developed nectariferous disc; 2-carpellate pistil; samaroid schizocarp splitting into 2 segments; and non-endospermous seeds. Anatomical features: mucilaginous leaf epidermis, secretory cells containing latex and/or other substances in leaf tissues (phloem and/or mesophyll, also sometimes in stem); tissues with calcium oxalate crystals; and trilacunar nodes.

Genera/species: 2/210

Distribution: Widespread primarily in the temperate zone of the Northern Hemisphere; center of diversity in China. Components of various deciduous and mixed forests.

Major genus: *Acer* (about 200 spp.)

Florida representatives: *Acer* (5 spp.)

Economic plants and products: Sugar and syrup from the sap of several species, especially *Acer saccharum* (sugar maple). Several timber and ornamental trees, including: *A. nigrum* (black maple), *A. palmatum* (Japanese maple), *A. rubrum* (red maple), *A. saccharinum* (silver maple), and *A. saccharum*.

46

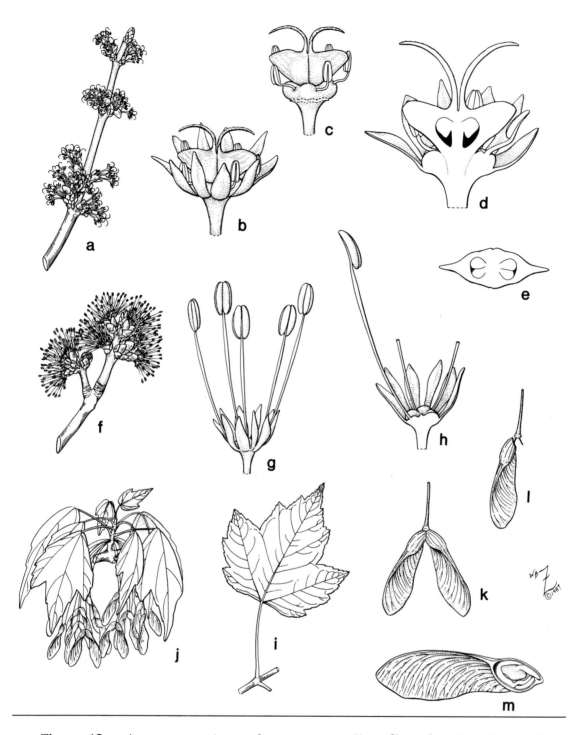

Figure 18. Aceraceae. *Acer rubrum:* **a,** carpellate flowering branch, x 4/5; **b,** carpellate flower, x 6; **c,** androecium and gynoecium of carpellate flower, x 6; **d,** longitudinal section of flower, x 10; **e,** cross section of ovary, x 10; **f,** staminate flowering branch, x 4/5; **g,** staminate flower, x 4 1/2; **h,** longitudinal section of staminate flower, x 6; **i,** mature leaf, x 1/2; **j,** fruiting branch, x 2/5; **k,** schizocarp, x 1; **l,** dehiscing schizocarp, x 1; **m,** longitudinal section of schizocarp segment, x 2.

The maples are extremely variable morphologically and even have been divided into two or more genera (Murray, 1970). General botanical opinion favors the retention of a single genus. *Acer* is divided into numerous distinct sections based upon morphological characters of the inflorescence, flowers, leaves, and fruit, as well as characters of the winter buds (Brizicky, 1963). The key characters used to delimit species are usually based upon morphology of the leaves, which may vary from simple (most species) to compound (e.g., *A. negundo*). The simple leaves, which have palmate venation, vary in the number of lobes and the type of leaf margin (entire to deeply toothed).

Maple flowers are usually imperfect, although some perfect flowers may occur within an inflorescence. In most species, the plants are basically monoecious with the staminate and carpellate inflorescences situated on different parts of the same individual and developing at different times. Usually the flowers, as in *A. rubrum*, are entomophilous. *A. negundo*, a notable exception, is a dioecious, wind-pollinated species. Insects of various types (often bees) are attracted to the copious and easily accessible nectar that is exposed on the disc at the base of the ovary. Cross-pollination is insured by the nonsynchonous flowering of the staminate and carpellate flowers (and the protogyny of the perfect flowers).

References Cited

Brizicky, G.K. 1963. The genera of Sapindales in the southeastern United States. J. Arnold Arbor. 44: 462–501.
Murray, A.E. 1970. A monograph of the Aceraceae. 332 pp. Ph.D. Dissertation. Pennsylvania State Univ., State College, PA.

MELASTOMATACEAE (MELASTOME OR MEADOW-BEAUTY FAMILY)

Annual or perennial herbs, shrubs, or trees, erect or climbing; stems square or round in cross section. Leaves simple, entire or toothed, opposite (one of pair sometimes smaller than the other), decussate, sessile to petiolate, with usually 3 to 5 prominent and subparallel veins diverging from the base and converging at the apex (connected by transverse cross-veins), exstipulate. Inflorescence determinate, cymose, terminal or axillary. Flowers slightly zygomorphic (due to orientation of androecium), perfect, epigynous or sometimes half-epigynous or perigynous, large and showy to minute, sometimes subtended by showy bracts, with well-developed and often urceolate (urn-shaped) hypanthium. Calyx of usually 4 or 5 sepals, often distinct, imbricate, persistent. Corolla of usually 4 or 5 petals, distinct, usually spreading, often red, blue, white, pink, or purple, usually imbricate and/or convolute. Androecium of twice as many stamens as petals, biseriate; filaments distinct, inflexed in bud, geniculate; anthers 2- or sometimes 1-loculed, basifixed, often with various appendages at base of connectives, each dehiscing by an apical pore. Gynoecium of 1 pistil, usually 3- to 5-carpellate; ovary inferior (adnate to hypanthium) or sometimes half-inferior or superior (within free hypanthium), with as many locules as carpels; ovules numerous, anatropous or sometimes campylotropous, placentation usually axile; style 1, simple; stigma 1, capitate or punctate. Fruit a loculicidal capsule or berry; seeds numerous, small; endosperm absent; embryo minute, straight or conforming to seed shape, usually with unequal cotyledons.

Family characterization in Florida: herbaceous or woody plants with opposite, simple, exstipulate leaves sometimes with prominent subparallel venation (connected by transverse cross-veins); 4- or 5-merous flowers with urceolate hypanthium; stamens with geniculate filaments and often appendaged anthers with apical pores; and a loculicidal capsule or berry as the fruit type. Tissues commonly with calcium oxalate crystals.

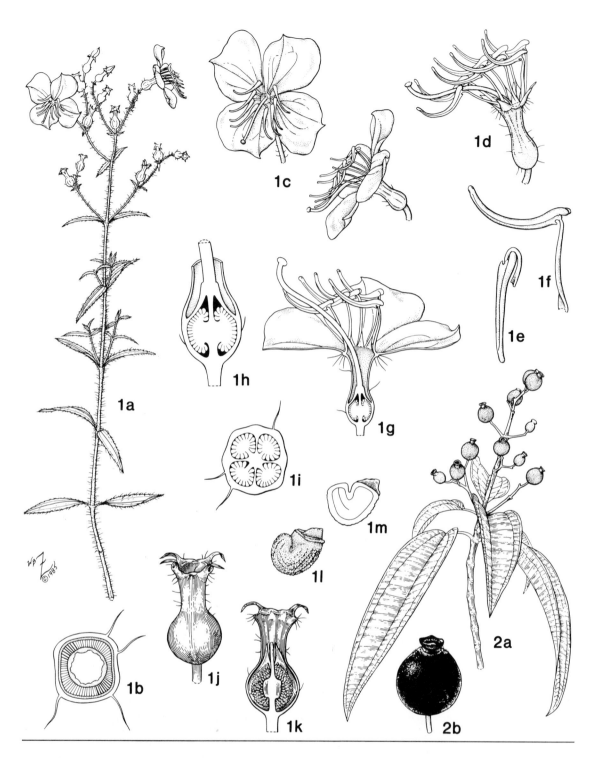

Figure 19. Melastomataceae. 1, *Rhexia mariana:* **a,** habit, x 1/2; **b,** cross section of stem showing intraxylary phloem, x 9; **c,** two views of flower, x 1; **d,** flower with petals removed, x 2; **e,** stamen from bud, x 4 1/2; **f,** stamen from mature flower, x 4 1/2; **g,** longitudinal section of flower, x 2; **h,** longitudinal section of ovary, x 6; **i,** cross section of ovary, x 6; **j,** fruit (capsule subtended by hypanthium), x 3; **k,** longitudinal section of fruit, x 3; **l,** seed, x 30; **m,** longitudinal section of seed, x 30. **2,** *Tetrazygia bicolor:* **a,** fruiting branch, x 1/2; **b,** berry, x 2.

Anatomical features: intraxylary phloem in the stem and often vascular bundles in the cortex and pith.

Genera/species: 244/3,360

Distribution: Pantropical; most diverse in South America.

Major genera: *Miconia* (1,000 spp.), *Medinilla* (400 spp.), *Memecylon* (300 spp.), *Tibouchina* (200+ spp.), *Leandra* (200 spp.), and *Clidemia* (150 spp.)

Florida representatives: *Rhexia* (10 spp.) and *Tetrazygia* (1 sp.); see Kral and Bostick (1969).

Economic plants and products: Lumber (used locally for furniture and construction) from *Astronia* and *Memecylon*. Edible fruits from: *Medinilla*, *Melastoma*, and *Mouriri*. A few dye plants, such as *Memecylon*. Ornamental plants (species of 19 genera), including: *Dissotis*, *Heterocentron*, *Medinilla*, *Melastoma*, *Rhexia* (meadow-beauty), and *Tibouchina* (princess-flower).

The Melastomataceae are a well-defined and natural family (Dahlgren and Thorne, 1984). The plants are easily recognized in the field by the distinctive leaf venation (not obvious in *Rhexia*) and stamens. The leaves (e.g., *Tetrazygia*) are often characterized by prominent secondary veins that diverge from the base and converge at the apex; these major veins and the midvein are connected by transverse tertiary veins. The stamens typically are geniculate (bent), dehisce by apical pores, and have modified appendages of the connective (Wilson, 1950). The morphology of these appendages (e.g., awl-shaped, spiny, curved) have been used to distinguish species and genera.

The showy flowers generally do not produce nectar, and insects (often bumblebees) visit to collect pollen. The flowering of *Rhexia* species overlaps in time and location, resulting in several natural hybrids (see Wurdack and Kral, 1982).

The distinctive fruit of *Rhexia* is a capsule enclosed within the lower globose portion of the persistent flask-shaped hypanthium. The fruit of *Tetrazygia* is a berry.

References Cited

Dahlgren, R.M.T. and R.F. Thorne. 1984. The order Myrtales: circumscription, variation, and relationships. Ann. Missouri Bot. Gard. 71: 633-699.

Kral, R. and P.E. Bostick. 1969. The genus *Rhexia* (Melastomataceae). Sida 3: 387-440.

Wilson, C.L. 1950. Vasculation of the stamen in the Melastomataceae, with some phyletic implications. Amer. J. Bot. 37: 431-444.

Wurdack, J.J. and R. Kral. 1982. The genera of Melastomataceae in the southeastern United States. J. Arnold Arbor. 63: 429-439.

ONAGRACEAE (EVENING-PRIMROSE FAMILY)

Annual, biennial, or usually perennial herbs or sometimes shrubs or trees, occasionally aquatic. <u>Leaves</u> simple, entire, toothed, or lobed, alternate, opposite, or whorled, stipulate (stipules minute and caducous) or exstipulate. <u>Inflorescence</u> indeterminate, spicate, racemose, paniculate, or flowers solitary and axillary. <u>Flowers</u> actinomorphic or sometimes zygomorphic, usually perfect, epigynous, often showy, with well-developed tubular to campanulate hypanthium (adnate to and often produced beyond the

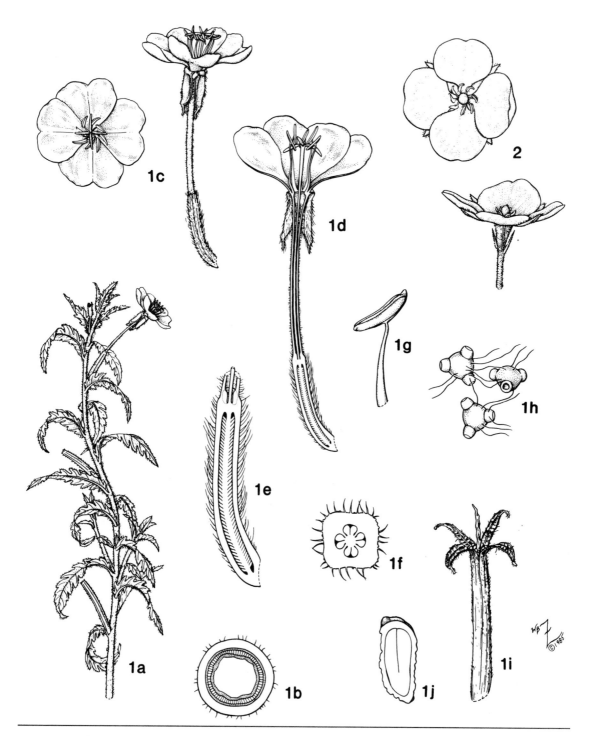

Figure 20. Onagraceae. 1, *Oenothera laciniata:* **a,** habit, x 1/2; **b,** cross section of stem showing intraxylary phloem, x 9; **c,** two views of flower, x 1 1/4; **d,** longitudinal section of flower, x 1 1/2; **e,** longitudinal section of ovary, x 3; **f,** cross section of ovary, x 9; **g,** anther, x 9; **h,** pollen grains connected by viscin strands, x 80; **i,** capsule, x 2; **j,** longitudinal section of seed, x 20. **2,** *Ludwigia peruviana:* two views of flower, x 3/4.

ovary), with nectariferous disc (at base of style) or nectaries (at lower part of hypanthium). <u>Calyx</u> of usually 4 sepals, distinct, deciduous or persistent, valvate. <u>Corolla</u> of usually 4 petals, distinct, often clawed, yellow, white, red, or orange, convolute, imbricate or valvate. <u>Androecium</u> of 8 (biseriate) or sometimes 4 (uniseriate) stamens, arising from or near hypanthium rim; filaments distinct; anthers versatile, basifixed, sometimes cross-partitioned, dehiscing longitudinally; pollen in tetrads or monads, connected by viscin strands, each grain with 3 pores and protruding "stoppers" (apertures). <u>Gynoecium</u> of 1 pistil, 4-carpellate; ovary inferior, 4-loculed but sometimes with septa incomplete at base; ovules several to numerous in each locule, anatropous, placentation axile; style 1, slender; stigma 1 and capitate, notched, or 4-lobed, or of 4 radiate branches. <u>Fruit</u> usually a loculicidal capsule or sometimes a berry or nutlet; seeds usually numerous, comose or glabrous; embryo oily, generally straight; endosperm absent.

Family characterization in Florida: perennial herbs or shrubs, 4-merous flowers with short to elongate hypanthium; distinctive pollen grains (with 3 protruding apertures) in monads or tetrads and connected by viscin strands (see Skvarla et al., 1975); and inferior ovary (often terminated by prolonged hypanthium). Tissues with raphides (bundles of needle-like calcium oxalate crystals). Anatomical feature: intraxylary phloem (internal phloem) in the stems.

Genera/species: 18/650

Distribution: Temperate and subtropical regions (especially in the New World); most diverse in the western United States and Mexico.

Major genera: *Epilobium* (215 spp.), *Fuchsia* (100 spp.), *Oenothera* (80 spp.), and *Ludwigia* (75 spp.)

Florida representatives: *Ludwigia* (24-27 spp.), *Oenothera* (9 spp.), and *Gaura* (3 spp.)

Economic plants and products: Ornamental plants (species of 12 genera), including: *Clarkia* (farewell-to-spring, godetia), *Fuchsia* (lady's-eardrops), *Gaura*, and *Oenothera* (evening-primrose).

The Onagraceae, a well-defined group, are divided into several tribes primarily on the basis of fruit characters (Raven, 1964; Dahlgren and Thorne, 1984). The embryology, cytology, and chemistry (Averett and Raven, 1984) of the family have been extensively studied, especially within the genus *Oenothera* (Cleland, 1972). The delimitations of genera and species are often difficult due to intergradation (for example, see Raven et al., 1979).

The attractive flowers are usually entomophilous (or sometimes bird-pollinated), and typically open in the evening for pollinators (bees and Lepidoptera). Nectar, which accumulates in the floral tube, is secreted by a disc at the base of the style or by nectaries within the lower part of the hypanthium. The pollen adheres to the insects by means of a sticky substance (viscin) that forms strands between the grains (or tetrads). Although cross-pollination is generally promoted by protandry, self-pollination is also very prevalent in the family.

References Cited

Averett, J.E. and P.H. Raven. 1984. Flavonoids of Onagraceae. Ann. Missouri Bot. Gard. 71: 30-34.

Cleland, R.E. 1972. *Oenothera*: cytogenetics and evolution. 370 pp. Academic Press, Inc., London, England.

Dahlgren, R.M.T. and R.F. Thorne. 1984. The order Myrtales: circumscription, variation, and relationships. Ann. Missouri Bot. Gard. 71: 633-699.

Raven, P.H. 1964. The generic subdivision of Onagraceae, tribe Onagreae. Brittonia 16: 276-288.

_____, W. Dietrich, and W. Stubbe. 1979. An outline of the systematics of *Oenothera* subsect. *Euoenothera* (Onagraceae). Syst. Bot. 4: 242-252.

Skvarla, J.J., P.H. Raven, and J. Praglowski. 1975. The evolution of pollen tetrads in Onagraceae. Amer. J. Bot. 62: 6-35.

MYRTACEAE (MYRTLE FAMILY)

Trees or shrubs, aromatic (due to ethereal oils). <u>Leaves</u> simple, entire, opposite or alternate, coriaceous, persistent, glandular-punctate, exstipulate. <u>Inflorescence</u> determinate, basically cymose and often appearing racemose, umbellate, or paniculate, or sometimes flower solitary and axillary. <u>Flowers</u> actinomorphic, perfect, epigynous or sometimes half-epigynous, usually subtended by 2 bracts, with well-developed hypanthium (often prolonged beyond the ovary), with nectariferous disc (on summit of ovary or lining the hypanthium). <u>Calyx</u> of 4 or 5 sepals (sometimes much reduced or absent), distinct or basally connate, imbricate or sometimes undivided in bud (then splitting irregularly at anthesis or deciduous as a calyptra). <u>Corolla</u> of 4 or 5 petals, distinct (or sometimes connivent and forming a calyptra or absent), usually white or red, imbricate. <u>Androecium</u> of numerous stamens; filaments distinct or sometimes basally connate into 4 or 5 fascicles opposite the petals, inflexed in bud; anthers dorsifixed, versatile, with connective apex often glandular, usually dehiscing longitudinally, introrse. <u>Gynoecium</u> of 1 pistil, usually 2- to 5-carpellate; ovary inferior or sometimes half-inferior, 1-loculed or usually with as many locules as carpels; ovules 2 to many in each locule, anatropous or campylotropous, pendulous, placentation basically parietal (1-loculed ovary) but often appearing axile due to coalescent intruded placentae; style 1, elongate; stigma 1, capitate. <u>Fruit</u> usually a berry or loculicidal capsule; endosperm scanty and starchy or absent; embryo more or less bent or spirally rolled, variously shaped.

Family characterization in Florida: trees or shrubs with aromatic parts; simple, coriaceous, glandular-punctate leaves; 4- or 5-merous perianth and numerous stamens; inferior ovary with axile or deeply intruding parietal placentae; and a berry or loculicidal capsule as the fruit type. Tissues with calcium oxalate crystals (solitary or clustered) and a high tannin content. Anatomical features: schizogenous secretory cavities and lysigenous glands containing ethereal oils (appearing on the leaves as translucent dots); intraxylary phloem in the petiole (bicollateral bundles) and stem; and unilacunar nodes.

Genera/species: 144/3,000

Distribution: Tropical and subtropical regions; diverse in Australia and tropical America.

Major genera: *Eugenia* (1,000 spp.), *Eucalyptus* (500 spp.), *Myrica* (500 spp.), and *Syzygium* (500 spp.)

Florida representatives: 8 genera/18 spp.; largest genera: *Eugenia* (5 or 6 spp.), *Psidium* (3 spp.), *Calyptranthes* (2 spp.), *Myrtus* (2 spp.), and *Syzygium* (2 spp.)

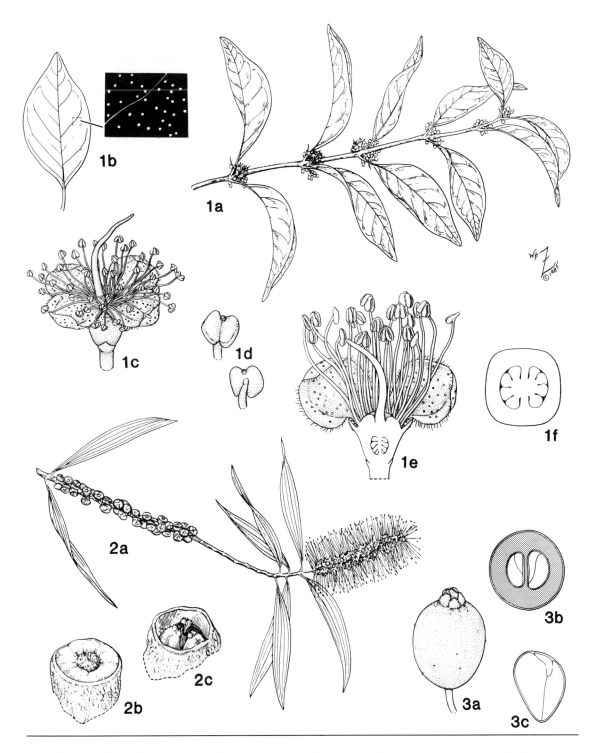

Figure 21. Myrtaceae. 1, *Eugenia axillaris:* **a,** flowering branch, x 1/2; **b,** leaf, x 1/2, with detail showing punctate surface, x 6; **c,** flower, x 8; **d,** two views of anther, x 25; **e,** longitudinal section of flower, x 9; **f,** cross section of ovary, x 20. **2,** *Melaleuca quinquenervia:* **a,** flowering and fruiting branch, x 1/2; **b,** fruit (capsule enclosed by hypanthium), x 4 1/2; **c,** dehisced fruit, x 4 1/2. **3,** *Myrcianthes fragrans:* **a,** berry, x 2; **b,** cross section of berry, x 2; **c,** longitudinal section of seed, x 3 1/2.

Economic plants and products: Edible fruits from: *Eugenia* (Surinam-cherry), *Myrciaria* (jaboticaba), *Psidium* (guava), and *Syzygium* (rose-apple). Spices and ethereal oils from: *Eucalyptus* (oil), *Melaleuca* (cajeput oil), *Pimenta* (allspice, the unripe berries; pimento; oil of bay rum), and *Syzygium* (cloves, the dried flower buds). Timbers from several, such as species of *Eucalyptus* and *Eugenia*. A weedy species in south Florida: *Melaleuca quinquenervia*. Ornamental trees and shrubs (species of 38 genera), including: *Callistemon* (bottle-brush), *Eucalyptus* (gumtree), *Eugenia* (Australian bush-cherry, Surinam-cherry), *Feijoa* (pineapple-guava), *Melaleuca* (bottle-brush, cajeput-tree, punk-tree), *Myrtus* (myrtle), and *Rhodomyrtus* (downy-myrtle).

The Myrtaceae are divided into two (or sometimes three) subfamilies based upon the fruit type (berry, capsule, or dry and indehiscent fruit; Dahlgren and Thorne, 1984). The generic delimitations and relationships vary considerably with different classifications (see Wilson, 1960; McVaugh, 1968; and Johnson and Briggs, 1984).

The prominent field characters are the glandular-punctate leaves, epigynous flowers, and numerous stamens. Insects (and sometimes birds) seeking nectar are attracted to the flowers, which may have very conspicuous stamens (e.g., *Melaleuca*, *Callistemon*), creating a "bottle-brush" effect, or showy petals with tufts of stamens as in *Myrtus* or *Myrcianthes*.

References Cited

Dahlgren, R.M.T. and R.F. Thorne. 1984. The order Myrtales: circumscription, variation, and relationships. Ann. Missouri Bot. Gard. 71: 633-699.

Johnson, L.A.S. and B.G. Briggs. 1984. Myrtales and Myrtaceae -- a phylogenetic analysis. Ann. Missouri Bot. Gard. 71: 700-756.

McVaugh, R. 1968. The genera of American Myrtaceae -- an interim report. Taxon 17: 354-418.

Wilson, K.A. 1960. The genera of Myrtaceae in the southeastern United States. J. Arnold Arbor. 41: 270-278.

OLEACEAE (OLIVE FAMILY)

Trees, shrubs, or sometimes lianas. Leaves simple or pinnately compound, usually entire, opposite, deciduous or persistent, often abaxially punctate, exstipulate. Inflorescence determinate, basically cymose and appearing racemose, paniculate, or fasciculate, terminal or axillary. Flowers actinomorphic, perfect or sometimes imperfect (then plants dioecious, polygamous, or polygamodioecious), hypogynous, small, with nectariferous disc often present around base of ovary. Calyx synsepalous with usually 4 lobes, valvate. Corolla usually sympetalous with 4 lobes or sometimes petals distinct or absent, variously colored, usually imbricate or valvate. Androecium of typically 2 stamens, epipetalous; filaments distinct, short; anthers basifixed, often apiculate, dehiscing longitudinally. Gynoecium of 1 pistil, 2-carpellate; ovary superior, 2-loculed; ovules usually 2 in each locule, anatropous or amphitropous, placentation axile; style 1 or absent; stigma 2-lobed or simple, capitate. Fruit a berry, drupe, loculicidal capsule, circumscissile capsule (pyxis), or samara; endosperm oily or absent; embryo straight, spatulate.

Family characterization in Florida: usually trees or shrubs with opposite, exstipulate and often abaxially punctate leaves; usually 4-merous perianth; 2 epipetalous stamens; and 2-loculed superior ovary. Tissues commonly with calcium oxalate crystals. Anatomical features: peltate trichomes (sometimes secretory) often on leaves and twigs (appearing on leaves as greenish or sunken dots) and one-trace unilacunar nodes.

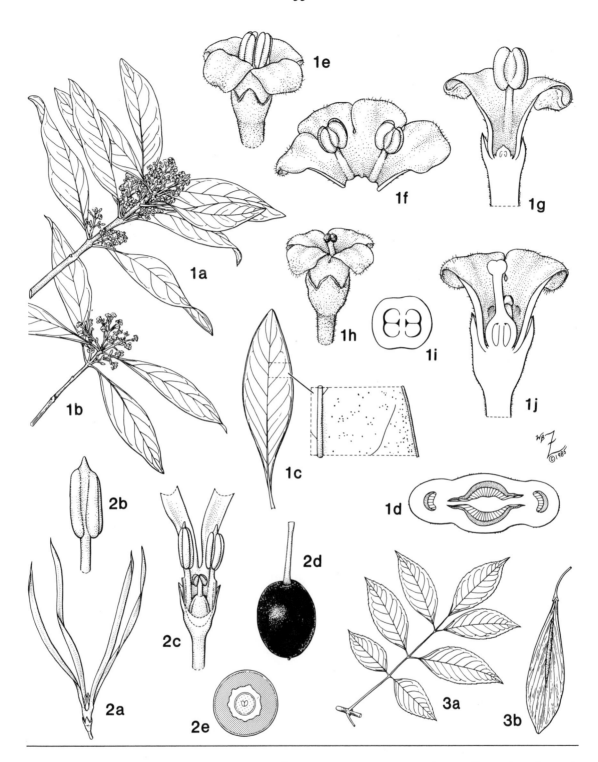

Figure 22. Oleaceae. 1, *Osmanthus americana:* **a,** staminate flowering branch, x 1/3; **b,** carpellate flowering branch, x 1/3; **c,** abaxial side of leaf, x 1/3, with detail showing punctate surface (peltate trichomes), x 2; **d,** cross section of node (one-trace unilacunar), x 4 1/2; **e,** staminate flower, x 6; **f,** expanded corolla and androecium of staminate flower, x 6; **g,** longitudinal section of staminate flower, x 7 1/2; **h,** carpellate flower, x 6; **i,** cross section of ovary, x 18; **j,** longitudinal section of carpellate flower, x 9. **2,** *Chionanthus virginica:* **a,** perfect flower, x 2; **b,** anther, x 10; **c,** detail of perfect flower (portion of perianth removed), x 7; **d,** drupe, x 1 1/2; **e,** cross section of drupe, x 1 1/2. **3,** *Fraxinus caroliniana:* **a,** leaf, x 1/3; **b,** samara, x 1 1/4.

Genera/species: 29/600

Distribution: Nearly cosmopolitan; particularly diverse in Asia and Australasia.

Major genera: *Jasminum* (300 spp.), *Chionanthus* (80-100 spp.), and *Fraxinus* (70 spp.)

Florida representatives: 6 genera/18 spp.; largest genera: *Forestiera* (4 spp.), *Fraxinus* (4 spp.), and *Jasminum* (4 or 5 spp.); see also Hardin (1974).

Economic plants and products: Fruit and oil from *Olea europaea* (olive). Perfume from several species of *Jasminum* (jasmine). Lumber from *Fraxinus* (ash). Ornamental trees and shrubs (species of 16 genera), including: *Chionanthus* (fringe-tree), *Forsythia* (golden-bells), *Jasminum*, *Ligustrum* (privet), *Osmanthus* (fragrant-olive), and *Syringa* (lilac).

The Oleaceae are generally considered a natural family, but the affinities to other groups are uncertain (see Wilson and Wood, 1959; Heywood, 1978; and Cronquist, 1981). The family is divided into two subfamilies (and several tribes) based upon the fruit type and ovule/seed position. For example, the fruit of *Fraxinus* is a dry, one-seeded samara, while those of *Osmanthus* (drupe) and *Ligustrum* (few-seeded berry) are baccate.

The small flowers, often clustered into dense inflorescences, are generally more or less tubular with spreading limbs. Various insects are attracted to the nectar (concealed at the corolla tube base) and often also to the strong sweet odor of the flowers. However, anemophily occurs in *Fraxinus* flowers (often imperfect), which have reduced (or absent) perianths.

References Cited

Cronquist, A. 1981. An integrated system of classification of flowering plants. Pp. 948-950. Columbia Univ. Press, New York, NY.

Hardin, J.W. 1974. Studies of the southeastern United States flora: 4. Oleaceae. Sida 5: 274-285.

Heywood, V.H. (ed.) 1978. Flowering plants of the world. Pp. 226-227. Mayflower Books, New York, NY.

Wilson, K.A. and C.E. Wood. 1959. The genera of Oleaceae in the southeastern United States. J. Arnold Arbor. 40: 369-384.

BIGNONIACEAE (BIGNONIA OR TRUMPET-VINE FAMILY)

Trees, shrubs, or most often woody vines (then with adventitious roots or tendrils). <u>Leaves</u> simple or more often palmately or pinnately compound with terminal leaflet often tendril-like, entire to serrate, opposite or occasionally whorled, decussate, often with glands at the petiole bases or in the leaf-vein axils, exstipulate. <u>Inflorescence</u> determinate, basically cymose and often appearing racemose, or flower solitary, terminal or axillary. <u>Flowers</u> zygomorphic, perfect, hypogynous, usually large and showy, with nectariferous disc. <u>Calyx</u> synsepalous with 5 lobes or teeth, sometimes truncate, campanulate or bilabiate. <u>Corolla</u> sympetalous with usually 5 lobes, campanulate, funnelform, to bilabiate, variously colored, imbricate. <u>Androecium</u> of typically 4 stamens and 1 staminode, epipetalous; filaments distinct; anthers basifixed, coherent in pairs or sometimes free, with divergent and often unequal locules (one seemingly above the other), dehiscing longitudinally, introrse. <u>Gynoecium</u> of 1 pistil, 2-carpellate; ovary

57

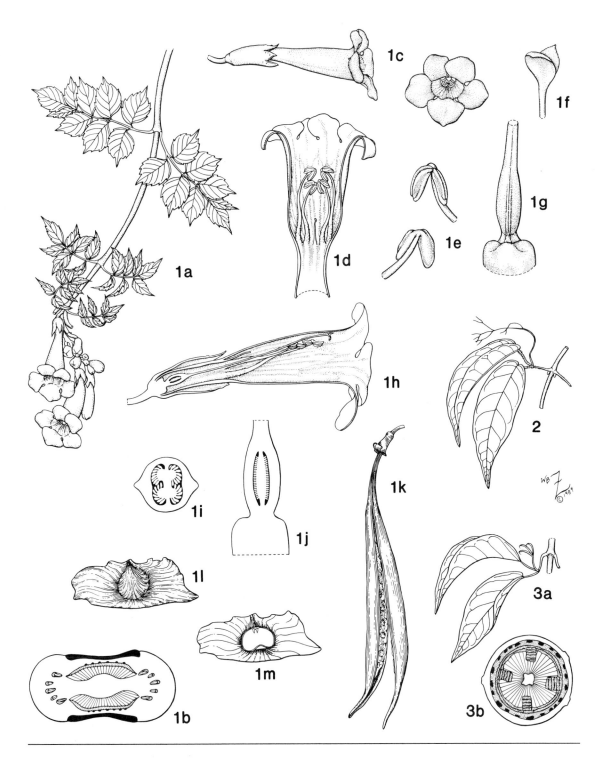

Figure 23. Bignoniaceae. 1, *Campsis radicans:* **a,** habit, x 1/3; **b,** cross section of node (one-trace multilacunar), x 6; **c,** two views of flower, x 1/2; **d,** expanded corolla and androecium, x 2/3; **e,** two views of anther, x 2; **f,** stigma, x 2; **g,** ovary and nectary, x 3; **h,** longitudinal section of flower, x 3/4; **i,** cross section of ovary, x 6; **j,** longitudinal section of ovary, x 3; **k,** capsule, x 1/2; **l,** seed, x 2 1/2; **m,** longitudinal section of seed, x 2 1/2. **2,** *Bignonia capreolata:* node, x 2/3. **3,** *Macfadyena unguis-cati:* **a,** node, x 1/2; **b,** cross section of young stem with secondary growth showing an anomalous pattern, x 9.

superior, 2-loculed or occasionally 1-loculed; ovules numerous, anatropous, erect, placentation axile (in 2-loculed ovary) or parietal (in 1-loculed ovary); style simple, filiform; stigma 1, 2-lobed (lobes flap-like). Fruit usually a 2-valved septicidal, loculicidal, or sometimes septifragal capsule, often woody, often elongate, or sometimes a berry or indehiscent pod; seeds numerous, compressed, winged, sometimes comose; endosperm typically absent; embryo straight.

Family characterization in Florida: trees or woody vines with opposite, usually compound leaves; zygomorphic, campanulate to bilabiate, showy corolla; androecium of 4 didynamous, epipetalous stamens and 1 staminode; divergent anther locules; 2-valved, woody, capsular fruit; and winged non-endospermous seeds. Tissues with calcium oxalate crystals. Anatomical features: anomalous secondary growth (usually involving the development of wedge-shaped masses of phloem in the xylem), unitegmic and tenuinucellate ovules, and 3- to several-trace unilacunar nodes.

Genera/species: 120/650

Distribution: Primarily tropical and subtropical with a few representatives in the temperate zone; particularly diverse in northern South America.

Major genera: *Tabebuia* (100 spp.) and *Jacaranda* (50 spp.)

Florida representatives: *Tecoma* (2 spp.), *Amphitecna* (1 sp.), *Bignonia* (1 sp.), *Campsis* (1 sp.), *Catalpa* (1 sp.), *Crescentia* (1 sp.), *Enallagma* (1 sp.), *Jacaranda* (1 sp.), *Macfadyena* (1 sp.), *Podranea* (1 sp.), *Pyrostegia* (1 sp.), and *Tabebuia* (1 sp.).

Economic plants and products: Lumber from *Tabebuia* (West Indian boxwood) and *Catalpa*. Ornamental trees and vines (species of 45 genera), including: *Bignonia* (crossvine, trumpet-flower), *Campsis* (trumpet-creeper, trumpet-vine), *Crescentia* (calabash-tree), *Clytostoma* (painted-trumpet), *Jacaranda*, *Kigelia* (sausage-tree), *Macfadyena* (cat-claw), *Pandorea*, *Paulownia* (empress-tree), *Pyrostegia* (flame-vine), *Spathodea* (flame-tree, African tulip-tree), *Tecoma* (yellow-elder), *Tabebuia* (gold-tree, roble blanco), and *Tecomaria* (Cape-honeysuckle).

The Bignoniaceae are divided into four to eight tribes based on characters of the ovary (2- or 1-loculed), fruit (various capsules or berry), and seeds (presence of wings). Specimens with both flowers and fruits are essential for critical determinations of genera and species (Gentry, 1973, 1980; Gentry and Tomb, 1979).

The majority of species are woody vines that are root climbers (*Campsis radicans*) or tendril climbers (most species, e.g., *Bignonia*). The tendrils, which represent modified leaflets, terminate the compound leaves. They may be branched and sometimes also end in hooks (*Macfadyena unguis-cati*).

Insects, hummingbirds, and bats visit the colorful, bell- or funnel-shaped flowers for the nectar secreted and concealed at the base. The chances of self-pollination are reduced by means of the "sensitive stigma," which consists of two flap-like lobes (Gentry, 1974). As a pollinator enters the flower, it first contacts the stigma and deposits pollen (from a previous flower) onto the inner surfaces of the spreading lobes. The sensitive lobes then close together. Later when the pollinator (covered with new pollen) backs away from that flower, it touches only the non-receptive outer surface of the stigmatic lobes.

References Cited

Gentry, A.H. 1973. Generic delimitations of Central American Bignoniaceae. Brittonia 25: 226-242.

_____. 1974. Coevolutionary patterns in Central American Bignoniaceae. Ann. Missouri Bot. Gard. 61: 728-759.

_____. 1980. Bignoniaceae -- Part 1. (Crescentieae and Tourrettieae). Fl. Neotropica Monogr. 25: 1-130.

_____ and A.S. Tomb. 1979. Taxonomic implications of Bignoniaceae palynology. Ann. Missouri Bot. Gard. 66: 756-777.

ACANTHACEAE (ACANTHUS FAMILY)

Perennial herbs, vines, or occasionally shrubs, sometimes armed. Leaves simple, usually entire, opposite, decussate, exstipulate. Inflorescence determinate, basically cymose and often appearing racemose, paniculate, or spicate (flowers congested in leaf axils), or sometimes flower solitary, axillary. Flowers zygomorphic, perfect, hypogynous, with nectariferous disc, often subtended by conspicuous bracts. Calyx synsepalous with 5 or sometimes 4 lobes, convolute or imbricate, persistent. Corolla sympetalous with 5 lobes, usually bilabiate, variously colored, imbricate or convolute. Androecium of usually 4 stamens and didynamous or sometimes reduced to 2, with staminodes often also present, epipetalous; filaments distinct or coherent in pairs; anthers more or less basifixed, 2- or 1-loculed, sometimes with locules unequal and at different levels, sometimes spurred and/or hairy, sometimes with prominent connective, dehiscing longitudinally. Gynoecium of 1 pistil, 2-carpellate; ovary superior, 2-loculed; ovules 2 to 10 in 2 rows in each locule, anatropous, amphitropous, or campylotropous, with modified funiculus (developed in fruit into a hook-shaped jaculator or retinaculum), placentation axile; style 1, slender, filiform; stigma(s) 1 or 2, funnelform or 2-lobed, often with reduced posterior lobe.. Fruit an elastic loculicidal capsule with 2 recurving valves; seeds usually flat, supported by small hook-like projections (retinacula or jaculators), with very thin, often mucilaginous (when moistened) seed coat; endosperm usually absent; embryo large, curved, bent, or sometimes straight.

Family characterization in Florida: herbs with opposite, simple leaves; bracteate flowers with usually bilabiate corollas; androecium of 4 (didynamous) or 2 stamens; anthers with unusual morphology (with hairs, spurs, expanded connective, and/or reduced to 1 locule); 2-valved, elastically dehiscing loculicidal capsule; and mucilaginous seeds with specialized hook-like funiculi (retinacula or jaculators). Tissues often with calcium oxalate crystals. Anatomical features: cystoliths in the parenchyma and epidermal cells in the leaves and stems (often appearing as raised streaks or lines); phloem with bundles of acicular fibers; and unitegmic and tenuinucellate ovules.

Genera/species: 256/2,770

Distribution: Pantropical with only a few representatives in temperate areas; centers of distribution in Indo-Malaya, Africa, Brazil, and Central America.

Major genera: _Justicia_ (300 spp.), _Ruellia_ (250 spp), _Strobilanthes_ (250 spp.), _Barleria_ (230 spp.), _Aphelandra_ (200 spp.), _Thunbergia_ (200 spp.), and _Dicliptera_ (150 spp.).

Florida representatives: 13-15 genera/24-29 spp.; largest genera: _Ruellia_ (4-7 spp.), _Justicia_ (4 spp.), _Dyschoriste_ (3 spp.).

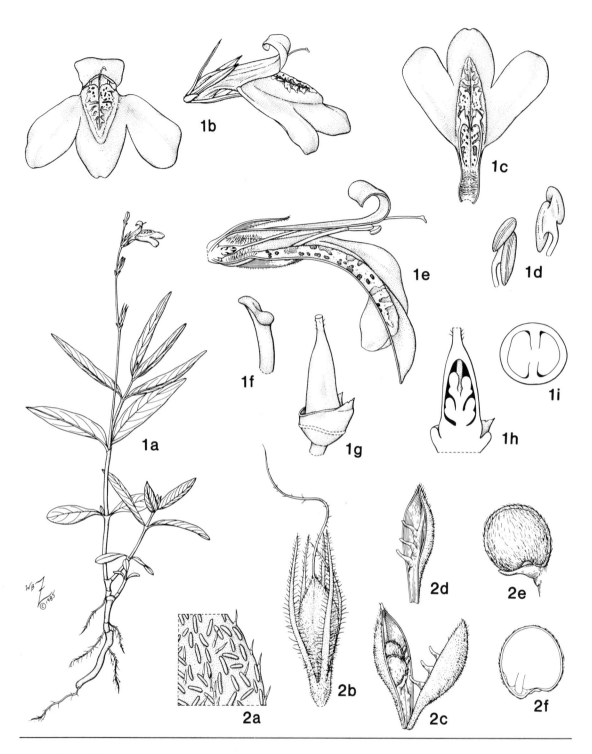

Figure 24. Acanthaceae. 1, *Justicia angusta* (*J. ovata* var. *angusta*): **a,** habit, x 1/2; **b,** two views of flower, x 2; **c,** lower corolla lip and androecium, x 2; **d,** two views of anther, x 6; **e,** longitudinal section of flower, x 3 1/2; **f,** stigma, x 25; **g,** ovary and nectary, x 10; **h,** longitudinal section of ovary, x 12; **i,** cross section of ovary, x 25. **2,** *Ruellia caroliniensis:* **a,** portion of dried leaf showing cystoliths, x 25; **b,** capsule before dehiscence, x 3; **c,** dehiscing capsule (persistent calyx removed), x 3; **d,** one valve of capsule showing retinacula (seeds ejected), x 3; **e,** seed (dry) with retinaculum, x 6; **f,** longitudinal section of seed, x 6.

Economic plants and products: A few dye plants, such as *Adhatoda* (yellow dye) and *Strobilanthes* (blue dye). Ornamental plants (species of 36 genera), including: *Acanthus* (bear's-breech), *Aphelandra* (zebra-plant), *Asystasia* (coromandel), *Beloperone* (shrimp-plant), *Crossandra*, *Eranthemum* (blue-sage), *Fittonia* (mosaic-plant), *Graptophyllum* (caricature-plant), *Justicia*, *Odontonema* (firespike), *Pachystachys* (cardinal's-guard, yellow shrimp-plant), *Pseuderanthemum*, *Sanchezia*, and *Strobilanthes*.

The Acanthaceae have been divided into three or four subfamilies with the majority of species placed in the Acanthoideae, a natural and well-defined group. The other subfamilies, sometimes segregated as families, have less modified funiculi and lack other characters, such as cystoliths. The phylogenetic relationships of the Acanthaceae are also controversial, with affinities to such groups as the Scrophulariaceae, Gesneriaceae, Verbenaceae, and Lamiaceae cited in the literature (see Mohan Ram and Wadhi, 1965 and Long, 1970 for summaries).

Genera are distinguished by characters of the bracts (size and form), corolla (shape), stamens (number and morphology), and ovules (number). Pollen morphology and sculpturing are also often used to delimit genera. Cystoliths, which often show up as streaks on the leaf blades, are characteristic of certain genera and tribes. In addition, numerous embryological studies have clarified the relationships of genera (and subfamilies).

The anthers, in particular, show a great variety of number, form, and position in different genera. For example, the flowers of *Thunbergia* have four stamens, while those of *Justicia* have only two. The anther locules, often at different levels, may also be unequal in size (or sometimes one is absent) and are often separated by a well-developed connective. In addition, the anthers are often spurred and/or hairy.

With brightly colored corollas, often showy bracts, and nectar produced by the discs, the flowers of the Acanthaceae attract insects (or sometimes birds). The flowers are protogynous. The corollas are generally either tubular with a five-lobed limb (*Ruellia*) or, more often, more or less bilabiate (*Justicia*) with an erect upper lip (bifid at apex) and a lower 3-lobed horizontal lip. The lower lip of the bilabiate flower serves as a landing platform for insect visitors. An insect becomes dusted with pollen as it enters and touches the anther locule tips (or spurs); it then transfers the pollen to the stigma of the next flower visited.

Most genera of the Acanthaceae are characterized by a bilocular capsule that splits elastically, leaving a persistent central column. The seeds are situated upon hook-shaped projections that represent modified funiculi (called retinacula or jaculators). As the valves of the capsule dehydrate and recurve, the jaculators (usually slightly twisted to one side) help to direct the seeds laterally.

References Cited

Long, R.W. 1970. The genera of Acanthaceae in the southeastern United States. J. Arnold Arbor. 51: 257-309.
Mohan Ram, H.Y. and M. Wadhi. 1965. Embryology and the delimitation of the Acanthaceae. Phytomorphology 15: 201-205.

VERBENACEAE (VERBENA FAMILY)

Annual or perennial herbs, shrubs, or trees, sometimes vines, often aromatic, often armed with prickles and/or thorns; stems and young twigs often square in cross section.

<u>Leaves</u> usually simple or sometimes palmately compound, entire or toothed, opposite or sometimes whorled, exstipulate. <u>Inflorescence</u> usually determinate, cymose and often appearing racemose, spicate or capitate, sometimes subtended by an involucre, terminal or axillary. <u>Flowers</u> zygomorphic, perfect, hypogynous, usually with inconspicuous nectariferous disc. <u>Calyx</u> synsepalous with 5 lobes or teeth, campanulate or tubular, aestivation open, persistent, sometimes expanded or enlarged in fruit. <u>Corolla</u> sympetalous with 5 lobes, salverform (with very narrow tube and abruptly spreading limb) or sometimes campanulate, often bilabiate, variously colored, imbricate. <u>Androecium</u> of 4 stamens and sometimes 1 staminode, didynamous, epipetalous; filaments distinct; anthers dorsifixed, dehiscing longitudinally, introrse. <u>Gynoecium</u> of 1 pistil, 2-carpellate; ovary superior, basically 2-loculed but often appearing 4-loculed due to false septa, usually slightly 4-lobed; ovules solitary in each apparent locule, anatropous or less often hemitropous or orthotropous, placentation axile; style 1, terminal or inserted in depression at ovary apex; stigma 1, lobed. <u>Fruit</u> a drupe with 2 or 4 pyrenes or sometimes a schizocarp splitting into 4 nutlets or single-seeded with a leathery wall (splitting when the embryo germinates); endosperm usually absent; embryo straight, oily.

Family characterization in Florida: aromatic herbs, shrubs, or trees often with square stems (and twigs) and opposite leaves; sympetalous, salverform, and bilabiate corolla; 4 epipetalous and didynamous stamens; 2-carpellate and 4-loculed ovary with 4 ovules; and a drupe (with 2 or 4 stones), schizocarp (splitting into 4 nutlets), or a leathery-skinned (1-seeded) fruit. Tissues often with calcium oxalate crystals. Anatomical features: xylem (even in herbs) forming a continuous ring with narrow rays, and unitegmic and tenuinucellate ovules.

Genera/species: 104/3,200

Distribution: Tropical and subtropical with only a limited number of representatives (usually herbs) in temperate regions.

Major genera: *Clerodendrum* (400 spp.), *Verbena* (250 spp.), *Vitex* (250 spp.), *Lippia* (220 spp.), *Premna* (200 spp.), and *Lantana* (150 spp.)

Florida representatives: 13-15 genera/37-39 spp.; largest genera: *Verbena* (10 spp.), *Lantana* (5-7 spp.), *Clerodendrum* (5 spp.), *Glandularia* (3 spp.), and *Lippia* (3 spp.)

Economic plants and products: Timber from *Tectona* (teak), *Citharexylum* (fiddlewood), and *Vitex*. Essential oils from *Lippia* (verbena oil). Herbal remedies from *Verbena* (vervain) and tea-leaves and edible tubers from *Priva*. Ornamental plants (species of 21 genera), including: *Callicarpa* (beauty-berry), *Clerodendrum* (bleeding-heart, glory-bower), *Congea* (woolly congea), *Duranta* (golden-dewdrop), *Holmskioldia* (Chinese hat-plant), *Lantana* (shrub-verbena), *Petrea* (queen's-wreath), *Verbena* (vervain), and *Vitex* (chaste-tree, monk's-pepper).

The Verbenaceae are divided into several subfamilies (and tribes) that are delimited according to the type of inflorescence and fruit (see Moldenke, 1977 and other papers of this series). A close affinity to the Lamiaceae is generally agreed upon by taxonomists (El-Gazzar and Watson, 1970). The family is especially diverse in habit and gynoecial morphology (see Cronquist, 1981), and several small groups are sometimes segregated as families, such as the Avicennioideae (Avicenniaceae), the black mangroves.

Bees and flies are attracted to the flowers and seek the nectar secreted at the base of the ovary. The flowers are generally protandrous (although *Avicennia* is protogynous), and pollination probably occurs in a similar pattern as in the mints (Lamiaceae).

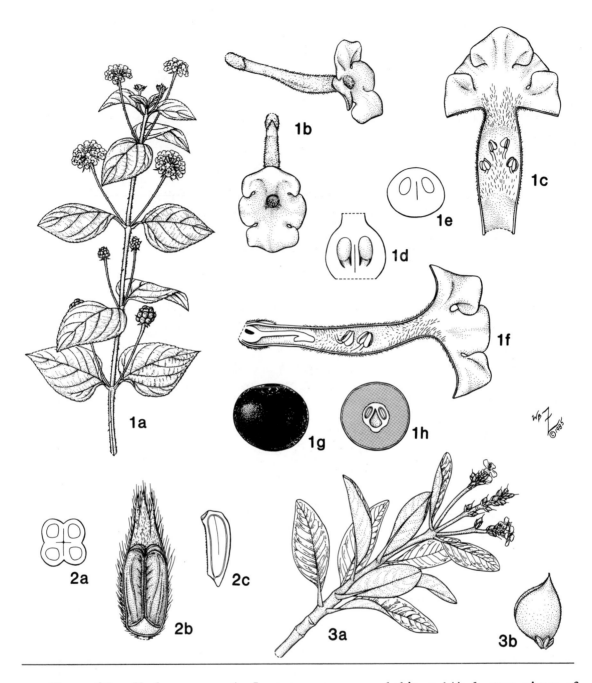

Figure 25. **Verbenaceae.** 1, *Lantana camara:* **a**, habit, x 1/4; **b**, two views of flower, x 3; **c**, expanded corolla and androecium, x 4 1/2; **d**, longitudinal section of ovary, x 18; **e**, cross section of ovary, x 18; **f**, longitudinal section of flower, x 6; **g**, drupe, x 3; **h**, cross section of drupe (note two-seeded pyrene), x 3. 2, *Verbena bonariensis:* **a**, cross section of ovary, x 36; **b**, schizocarp (half of persistent calyx removed), x 12; **c**, longitudinal section of schizocarp segment, x 12. 3, *Avicennia germinans:* **a**, flowering branch, x 1/2; **b**, fruit, x 3/4.

The fruit is often drupaceous, although a schizocarp (splitting into four nutlets) occurs in *Verbena* and *Glandularia*. The fruit of *Avicennia*, which is thin-walled and contains a single seed, splits as the embryo germinates. Genera with drupes are generally characterized by the arrangement of the pyrenes. For example, the drupe of *Vitex* has one four-chambered stone, while four distinct stones occur in *Clerodendrum*. The drupes of most *Lantana* species are characterized by two stones, although *L. camara* has one two-chambered stone.

References Cited

Cronquist, A. 1981. An integrated system of classification of flowering plants. Pp. 920-924. Columbia Univ. Press, New York, NY.

El-Gazzar, A. and L. Watson. 1970. A taxonomic study of Labiatae and related genera. New Phytol. 69: 451-486.

Moldenke, H.N. 1977. A fifth summary of the Verbenaceae, Avicenniaceae, Stilbaceae, Dicrastylidaceae, Symphoremaceae, Nyctanthaceae, and Eriocaulaceae of the world as to valid taxa, geographic distribution, and synonymy: supplement 7. Phytologia 36: 28-48.

BORAGINACEAE (BORAGE FAMILY)

Annual or perennial herbs commonly with rhizomes or taproots, or sometimes shrubs or trees, usually scabrous or hispid. <u>Leaves</u> simple, entire, alternate (lower ones sometimes opposite), exstipulate. <u>Inflorescence</u> determinate, basically a coiled cyme (helicoid or scorpioid cyme) and often appearing racemose or spicate, with the coiled axis bearing flowers along the "upper" side and straightening as the flowers mature, terminal. <u>Flowers</u> actinomorphic to slightly zygomorphic, usually perfect, hypogynous, generally with nectariferous disc. <u>Calyx</u> of 5 sepals, distinct or basally connate, generally campanulate, usually imbricate, often persistent. <u>Corolla</u> sympetalous with 5 lobes, rotate, salverform, funnelform, or campanulate, often with projecting infoldings or appendages (scales) in the throat, commonly blue, white, pink or yellow, imbricate or contorted. <u>Androecium</u> of 5 stamens, epipetalous; anthers basifixed or basally dorsifixed, dehiscing longitudinally, introrse. <u>Gynoecium</u> of 1 pistil, 2-carpellate; ovary superior, 2-loculed but becoming 4-loculed due to false septa, often deeply 4-lobed; ovules 4 (1 in each ovary section), anatropous to hemitropous, placentation axile but often appearing basal; style(s) 1 or 2, gynobasic or terminal; stigma(s) 1 (2-lobed) or 2, usually capitate, papillate. <u>Fruit</u> a schizocarp splitting into basically 4 nutlets or sometimes a 1- to 4-seeded drupe; endosperm scanty and fleshy or absent; embryo spatulate, erect, straight or sometimes curved.

Family characterization in Florida: often scabrous or hispid herbs; circinate and one-sided cymose inflorescence with the axis uncoiling as the flowers mature; corolla tube often with projecting infoldings or scales (a "corona"); deeply 4-lobed to unlobed, 2-carpellate ovary with gynobasic or terminal style; a schizocarp (splitting into 4 nutlets) or drupe as the fruit type; and seeds with little or no endosperm. Parenchymatous tissues commonly with calcium oxalate crystals. Anatomical features: firm, unicellular hairs with thick, often calcified and silicified walls and basal cystoliths; xylem (even in herbs) forming a continuous ring with narrow rays; and unitegmic and tenuinucellate ovules.

Genera/species: 114/2,400

Distribution: Cosmopolitan, but particularly well-represented in temperate and subtropical regions; centers of diversity in the Mediterranean region and western North America.

Major genera: *Cordia* (250 spp.), *Heliotropium* (250 spp.), *Tournefortia* (150 spp.), *Onosma* (150 spp.), and *Cryptantha* (100 spp.)

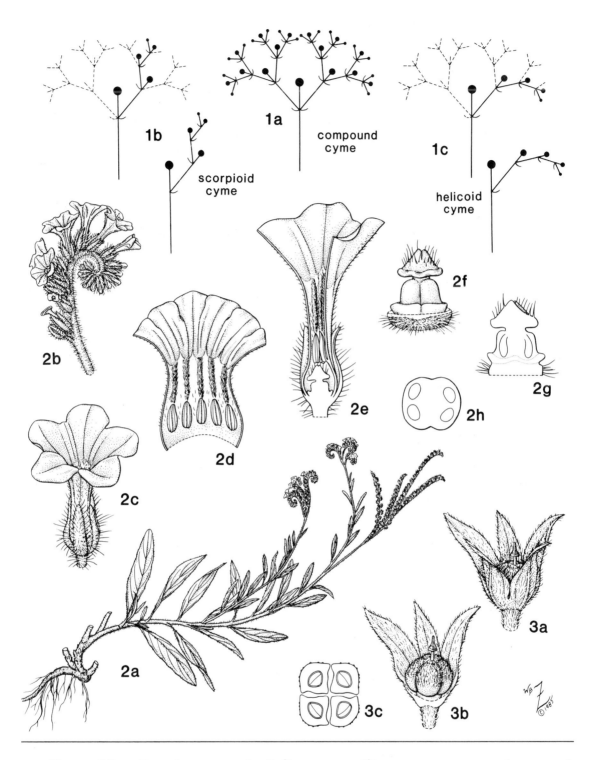

Figure 26. Boraginaceae. 1, Inflorescence diagrams: **a,** compound cyme; **b,** scorpioid cyme; **c,** helicoid cyme. **2,** *Heliotropium amplexicaule:* **a,** habit, x 1/3; **b,** inflorescence, x 2; **c,** flower, x 4 1/2; **d,** expanded corolla and androecium, x 6; **e,** longitudinal section of flower, x 8; **f,** pistil and nectary, x 15; **g,** longitudinal section of pistil, x 15; **h,** cross section of ovary, x 25. **3,** *Heliotropium polyphyllum:* **a,** schizocarp with persistent calyx, x 6; **b,** schizocarp (portion of calyx removed), x 6; **c,** cross section of schizocarp, x 10.

Florida representatives: 10 genera/23 spp.; largest genera: *Heliotropium* (7 spp.), *Bourreria* (3 spp.), and *Lithospermum* (3 spp.); see Ward and Fantz (1977).

Economic plants and products: Timber from *Cordia* spp. Alkanet (a red dye) from roots of *Alkanna*. Medicinal herbs: *Borago* (borage), *Lithospermum* (puccoon -- used as a contraceptive by certain Indians of North America), and *Symphytum* (comfrey). Ornamental plants (species of 32 genera), including: *Borago*, *Cerinthe* (honeywort), *Cordia* (geiger-tree), *Cynoglossum* (hound's-tongue), *Echium* (viper's-bugloss), *Heliotropium* (heliotrope), *Mertensia* (Virginia bluebells), *Myosotis* (forget-me-nots), and *Pulmonaria* (lungwort).

The Boraginaceae are generally divided into two to five subfamilies based upon characters of the style (simple or bilobed, terminal or gynobasic) and fruit (schizocarp or drupe). Mature fruits are very essential in most treatments for the identification of genera.

The characteristic dorsiventral and coiled inflorescences are very complex and diverse assortments of basically cymose arrangements. The major axes are sympodial, producing circinate cymes that uncoil progressively as the flowers open. The borage inflorescence type has been generally termed "helicoid" and/or "scorpioid" by various authors (see Rickett, 1955). The coiling is caused by the development of only one flower (or branch) of each original lateral pair of the cyme. Generally, in a helicoid cyme, the lateral branches develop from the same side of the axis, and in a scorpioid cyme, on alternating sides (see Fig. 26, 1a-1c). A seemingly spicate or racemose arrangement results.

Various insects (bees, Lepidoptera, flies) visit the flowers for the nectar secreted at the base. Infoldings, appendages, and/or hairs of the corolla aid in the protection and partial concealment of the nectar. Self-pollination is evidently prevalent in species with relatively inconspicuous flowers (Knuth, 1909).

References Cited

Knuth, P. 1909. Handbook of flower pollination. Trans. by J.R.A. Davis. Vol. III. Pp. 115-142. Oxford Univ. Press, Oxford, England.

Rickett, H.W. 1955. Materials for a dictionary of botanical terms. III. Inflorescences. Bull. Torrey Bot. Club 82: 419-445.

Ward, D.B. and P.R. Fantz. 1977. Keys to the flora of Florida -- 3, Boraginaceae. Phytologia 36: 309-323.

CONVOLVULACEAE (MORNING-GLORY FAMILY)

Annual or perennial herbs or sometimes shrubs or small trees, usually twining and climbing or prostrate, commonly with milky sap, with rhizomes or tuberous roots or stems. Leaves simple, entire, lobed, or pinnately divided to pectinate, alternate, exstipulate. Inflorescence determinate, cymose, or flowers solitary, axillary, with jointed peduncles. Flowers actinomorphic, perfect, hypogynous, often large and showy, ephemeral, usually with intrastaminal disc, generally subtended by a pair of bracts (sometimes enlarged and forming an involucre). Calyx of 5 sepals, distinct or sometimes basally connate, sometimes unequal, imbricate, persistent. Corolla sympetalous, entire to slightly 5-lobed, funnelform or salverform, plicated, brightly colored (commonly red, violet, blue, or white), induplicate-valvate and/or convolute (twisted) in bud. Androecium of 5 stamens, epipetalous at corolla base; filaments distinct, often unequal; anthers dorsifixed, dehiscing longitudinally, usually introrse. Gynoecium of 1 pistil, 2-

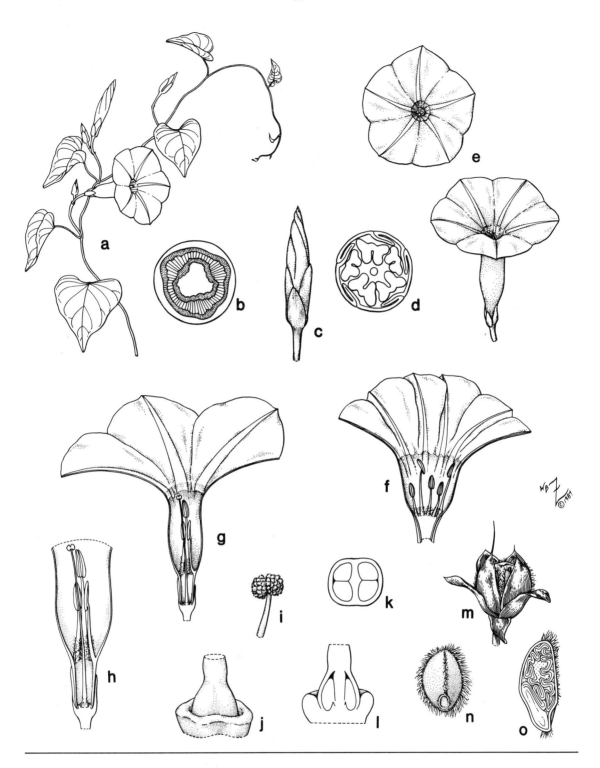

Figure 27. Convolvulaceae. *Ipomoea pandurata:* **a,** habit, x 1/3; **b,** cross section of stem showing intraxylary phloem, x 15; **c,** bud, x 1; **d,** cross section of bud (induplicate-valvate and convolute aestivation), x 6; **e,** two views of flower, x 1/2; **f,** expanded corolla and androecium, x 1/2; **g,** longitudinal section of flower, x 3/4; **h,** longitudinal section of flower (apical portion of corolla removed), x 1; **i,** stigma, x 6; **j,** ovary with nectariferous disc, x 6; **k,** cross section of ovary, x 9; **l,** longitudinal section of ovary, x 6; **m,** capsule, x 1 1/2; **n,** seed, x 3; **o,** longitudinal section of seed, x 4 1/2.

carpellate; ovary superior, 2-loculed or sometimes appearing 4-loculed due to false septa, sometimes with dense covering of hairs; ovules 2 in each locule, anatropous, sessile, placentation basal or basal-axile; style simple and filiform or forked; stigma(s) 1 or 2, linear, lobed or capitate. <u>Fruit</u> usually a 4-valved septifragal capsule; seeds smooth or hairy; endosperm scanty, hard, cartilaginous; embryo large, straight or curved, with folded or coiled, emarginate to bifid cotyledons, surrounded by endosperm.

Family characterization in Florida: climbing or prostrate, often twining herbs with milky sap; showy, actinomorphic, funnelform to salverform, plicated corolla with induplicate valvate and/or convolute aestivation; 5 epipetalous stamens; 2-carpellate ovary with axile placentation; septifragal capsule; and large embryo with folded, often bifid cotyledons. Tissues commonly with calcium oxalate crystals. Anatomical features: articulated latex canals or latex cells; intraxylary phloem in the petiole (bicollateral bundles) and stem; and unitegmic, generally tenuinucellate ovules.

Genera/species: 59/1,830

Distribution: Primarily in the tropics and subtropics with representatives having ranges extending into north and south temperate regions; particularly abundant in tropical America and tropical Asia.

Major genera: *Ipomoea* (500 spp.), *Convolvulus* (250 spp.), *Cuscuta* (170 spp.), and *Jacquemontia* (120 spp.)

Florida representatives: 14 genera/61 spp.; largest genera: *Ipomoea* (26 spp.; Austin, 1984), *Cuscuta* (8 or 9 spp.), *Stylisma* (5 spp.), *Jacquemontia* (5 spp.), and *Evolvulus* (4 spp.)

Economic plants and products: Edible tubers from *Ipomoea batatas* (sweet-potatoes, "yams"). Powerful drugs from several, such as species of *Convolvulus* (scammony, a purgative from the tuber) and *Ipomoea* (jalap, a purgative from the tubers, and lysergic acid, a hallucinogen from the seeds). Several weedy plants, such as *Convolvulus* (bindweed) and *Cuscuta* (dodder). Ornamental plants (species of 13 genera), including: *Calonyction* (moon-flower), *Calystegia* (bindweed), *Convolvulus*, *Dichondra*, *Ipomoea* (morning-glory, cypress-vine), *Porana* (Christmas-vine), and *Stylisma*.

The Convolvulaceae have been divided into three or four subfamilies (sometimes segregated as distinct families) and/or three to ten tribes. Although the relationships between these groups have been generally determined, the taxonomic rank (family, sub-family, or tribe) is a matter of controversy (see Wilson, 1960). A notable segregate group, the Cuscutoideae or Cuscutaceae (a monotypic taxon), has been separated from the rest of the Convolvulaceae by some botanists on the basis of the parasitic habit with related specializations of the corolla and embryo.

Authors also disagree on the delimitation of the various genera within the family. The generic lines depend upon characters of the bracts, sepals, corolla, pollen, stigma(s), and fruit. For example, the sepals vary in size, shape and pubescence, and the stigmas may be simple, lobed or globose. In addition, seed characters (e.g., type of pubescence) are important for species delimitation.

Morning-glories are easy to spot in the field with their twining habit and generally large, white or brightly colored, and funnel-shaped corolla. The corollas are twisted clockwise in bud and strongly plicated. Usually a flower is open for only one day (for a few hours); the corolla then incurves as it wilts. The corolla is characteristically divided longitudinally by five obvious demarcations that occur along the middle of the five lobes of the limb (see Fig. 27, e). These markings taper toward the apex and usually twist in the clockwise direction.

The flowers attract various insects (and in species of *Ipomoea*, birds), which visit for the nectar secreted on the hypogynous disc. The stamens closely surround the style by forming a short column in the center of the flower, and five narrow passages between the filament bases lead to the nectar. The insect may touch the protruding stigma as it enters the flower and then becomes dusted with pollen from the introrse anthers as it reaches for the nectar near the base. Self-pollination may occur when the flower wilts.

References Cited

Austin, D.F. 1984. Studies of the Florida Convolvulaceae -- IV. *Ipomoea*. Florida Scientist 47:81-87.
Wilson, K.A. 1960. The genera of Convolvulaceae in the southeastern United States. J. Arnold Arbor. 41: 298-317.

CAMPANULACEAE (BELLFLOWER FAMILY)

Annual or perennial herbs or sometimes shrubs, with milky or watery sap. <u>Leaves</u> simple, entire to pinnately divided, alternate, exstipulate. <u>Inflorescence</u> determinate, basically cymose and appearing racemose, paniculate, or sometimes capitate, terminal, or flowers solitary in leaf axils. <u>Flowers</u> actinomorphic or zygomorphic, usually perfect, epigynous or sometimes half-epigynous, generally showy, with nectariferous disc, sometimes resupinate (pedicel twisted 180 degrees during development), sometimes cleistogamous. <u>Calyx</u> synsepalous with typically 5 lobes, imbricate or valvate, often persistent. <u>Corolla</u> sympetalous with typically 5 lobes, campanulate, tubular, to bilabiate (then often split down one side), blue, red, violet, or sometimes white, valvate. <u>Androecium</u> of usually 5 stamens, epipetalous or attached to disc, sometimes unequal in length; filaments distinct to connate, often expanded basally (and forming a chamber over the disc; anthers basifixed, distinct, coherent, or connate (syngenesious), forming a cylinder around the style (even when anthers distinct), dehiscing longitudinally, introrse. <u>Gynoecium</u> of 1 pistil, typically 2-, 3-, or 5-carpellate; ovary usually inferior or sometimes half-inferior, usually with as many locules as carpels, lobed, commonly capped by nectariferous disc; ovules numerous, anatropous, placentation axile or occasionally parietal; style 1, sometimes 2-, 3-, or 5-branched, with pollen-collecting hairs below the stigmas; stigmas 2, 3, or 5, globose to cylindrical. <u>Fruit</u> a poricidal capsule with apical or basal pores and opening with flaps or slits or sometimes a circumscissile capsule or a berry; seeds numerous, small; endosperm copious, fleshy, oily; embryo small, straight, short to spatulate.

Family characterization in Florida: herbs to subshrubs often with milky sap; epigynous, zygomorphic or actinomorphic flowers with 5-merous perianth and androecium; coherent to connate anthers forming a cylinder around the style; numerous ovules on axile placentae; styles with pollen-collecting hairs; and a poricidal or circumscissile capsule as the fruit type. Tissues commonly with calcium oxalate crystals. Inulin (unusual storage polysaccharide) present. Anatomical features: articulated laticifers in phloem of stem and leaves, cystoliths often in leaf epidermal cells, often calcification or silicification of leaf epidermal cell walls, and unitegmic ovules.

Genera/species: 66/2,000

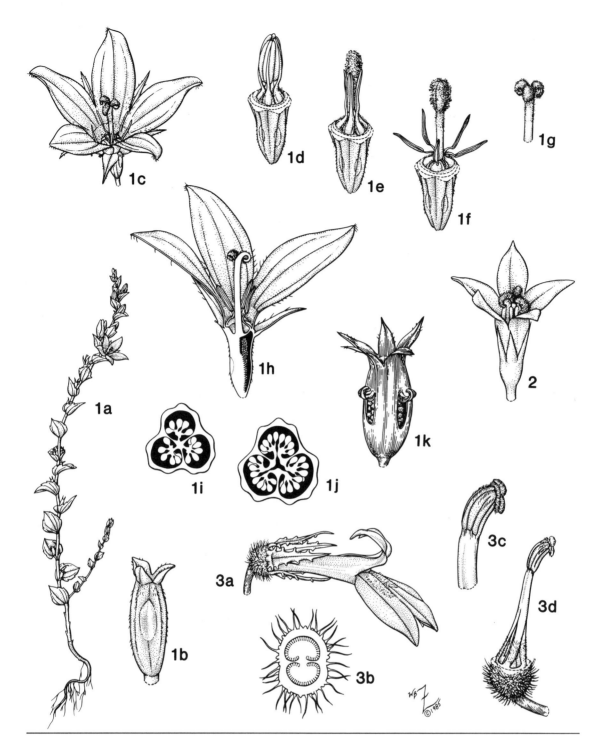

Figure 28. Campanulaceae. 1, *Triodanis perfoliata:* **a,** habit, x 1/2; **b,** cleistogamous flower, x 4 1/2; **c,** flower, x 3; **d,** androecium and gynoecium from a bud, x 4 1/2; **e,** androecium and gynoecium at anthesis (anthers dehisced), x 4 1/2; **f,** androecium and gynoecium of older flower (pollen deposited on pollen-collecting hairs of style branches), x 4 1/2; **g,** mature, expanded stigma, x 4 1/2; **h,** longitudinal section of flower, x 4; **i,** cross section of ovary at base, x 12; **j,** cross section of ovary near apex, x 12; **k,** capsule, x 6. **2,** *Wahlenbergia marginata:* flower, x 4 1/2. **3,** *Lobelia glandulosa:* **a,** flower, x 2; **b,** cross section of ovary, x 5; **c,** androecium and gynoecium, x 3; **d,** detail of anthers (syngenesious) and stigma, x 5.

Distribution: Generally widely distributed, especially in temperate and subtropical regions and the montane tropics.

Major genera: *Campanula* (300 spp.), *Lobelia* (200-300 spp.), *Centropogon* (230 spp.), and *Siphocampylus* (215 spp.)

Florida representatives: *Lobelia* (11 spp.), *Campanula* (2 spp.), *Triodanis* (2 sp.), *Wahlenbergia* (2 sp.), and *Sphenoclea* (1 sp.).

Economic plants and products: Several poisonous and medicinal plants (due to alkaloids), such as species of *Laurentia* (isotomin, a heart poison) and *Lobelia* (lobeline, a narcotic). Ornamental plants (species of 25 genera), including: *Campanula* (bellflower, harebell, bluebell), *Codonopsis* (bonnet-bellflower), *Edraianthus* (glassy-bells), *Jasione* (sheep's-bit), *Lobelia* (cardinal-flower), *Phyteuma* (horned rampions), *Triodanis* (Venus' looking-glass), and *Wahlenbergia*.

The Campanulaceae are often divided into three or four subfamilies, and some authors treat these groups as segregate families. The Lobelioideae, for example, are sometimes separated into their own family (Lobeliaceae) due to the zygomorphic and resupinate flowers, connate anthers, and also distinctive alkaloids. Genera within the Campanulaceae as a whole are often distinguished by fruit morphology, especially the mode of dehiscence (e.g., capsules with apical or basal pores).

The colorful flowers of the Campanulaceae attract various insects (especially bees), and in some cases, hummingbirds as the pollinators. Although the flowers of the family vary from open (*Triodanis*) to tubular and bilabiate (*Lobelia*), they share a basic pollen presentation mechanism similar to that of the Asteraceae (Shetler, 1979). The expanded filament bases fit closely together around the nectariferous disc at the base of the style, allowing only the insertion of a proboscis between them. The stamens, with filaments and/or anthers coherent to connate, form a tube around the style and the appressed, unexposed stigmas. Before or shortly after anthesis, the introrse anthers dehisce and fill the tube with pollen, which clings to the copious hairs often present on the style. As the "bottle-brush" style elongates, the pollen is pushed out and exposed to pollinators. Later the stigmas separate to expose the receptive upper surfaces. In some species, self-pollination may result when the stigmas recurve far enough to pick up some pollen still adhering to their own style (see Fig. 28, 1d-1g).

References Cited

Shetler, S.G. 1979. Pollen-collecting hairs of Campanula (Campanulaceae), I: historical review. Taxon 28: 205-215.

RHIZOPHORACEAE (RED MANGROVE FAMILY)

Shrubs or trees, frequently having a "mangrove habit" (with conspicuous prop roots) and growing in flooded (tidal) swamps; branches with swollen nodes. Leaves simple, entire, usually opposite, coriaceous, persistent; stipules usually conspicuous, interpetiolar, caducous, sometimes with colleters at inner surface of base. Inflorescence determinate, cymose or less often appearing racemose, axillary. Flowers actinomorphic, usually perfect, hypogynous to more commonly epigynous, with hypanthium (in epigynous flowers) sometimes prolonged beyond the ovary. Calyx of typically 4 or 5 to sometimes many sepals, usually basally connate, thick, usually fleshy or coriaceous, valvate, persistent. Corolla with as many petals as sepals, distinct, often clawed and lacerate or emarginate, usually fleshy or coriaceous, white or yellowish white, convolute or inflexed

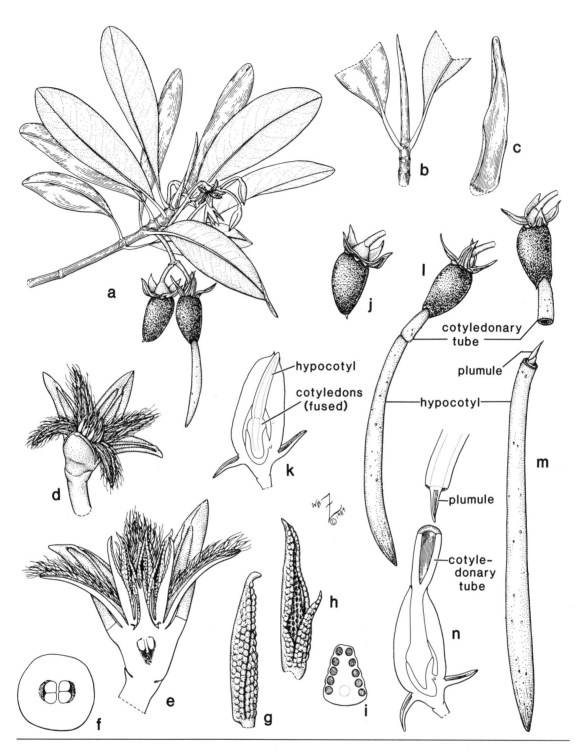

Figure 29. Rhizophoraceae. *Rhizophora mangle:* **a,** flowering and fruiting branch, x 1/2; **b,** apex of branch showing interpetiolar stipules, x 1/2; **c,** one stipule, x 1 1/4; **d,** flower, x 3; **e,** longitudinal section of flower, x 4 1/2; **f,** cross section of ovary, x 6; **g,** stamen (anther sessile), x 9; **h,** dehiscing anther, x 9; **i,** cross section of anther, x 15; **j,** fruit (seed in early stages of germination), x 2/3; **k,** longitudinal section of fruit, x 1; **l,** fruit with developing cotyledonary tube and hypocotyl, x 2/3; **m,** fruit and detached hypocotyl, x 2/3; **n,** longitudinal section of fruit and detached hypocotyl, x 1.

in bud. <u>Androecium</u> of usually 8 or 10 to sometimes many stamens, uniseriate, often in pairs opposite the petals; filaments distinct or connate at base, very short, or absent; anthers basifixed, 4-loculed or cross-partitioned, dehiscing longitudinally or by a flap, introrse. <u>Gynoecium</u> of 1 pistil, typically 2- to 4-carpellate; ovary superior, half-inferior to inferior, usually 2- to 4-loculed; ovules 2 in each locule, anatropous to hemitropous, pendulous, placentation axile or axile-basal; style 1; stigma 1, lobed. <u>Fruit</u> usually a leathery berry, drupe, or sometimes dry and indehiscent, terminated by the persistent calyx; seeds often viviparous (in mangrove species); endosperm copious or scanty, fleshy, oily, sometimes forming aril-like outgrowth; embryo straight, linear, often chlorophyllous, with enlarged hypocotyl.

Family characterization in Florida: woody plants with mangrove habit (prop roots) growing in muddy tidal shores, brackish streams, or lagoons; opposite, abaxially punctate, coriaceous leaves with large interpetiolar stipules; leathery 4-merous flowers; 8 uniseriate stamens in pairs opposite the adaxially hairy petals; sessile and chambered anthers, each dehiscing by a flap; leathery berry containing 1 seed (due to abortion); and viviparous and chlorophyllous embryo with enlarged hypocotyl. Tissues (especially the bark) with tannins and calcium oxalate crystals (solitary or clustered). Anatomical feature: trilacunar nodes.

Genera/species: 16/120

Distribution: Tropical to subtropical rain forests, shorelines, and muddy tidal flats; several genera composing a major component of the world's mangrove vegetation.

Major genus: *Cassipourea* (80 spp.)

Florida representative: *Rhizophora mangle*

Economic plants and products: Tannin (from bark and foliage) and wood (for charcoal and underwater pilings) from *Rhizophora* and several other genera. *Anopyxis* and *Rhizophora* (red mangrove) sometimes cultivated.

The taxonomic placement of the Rhizophoraceae presents a difficult problem (see Cronquist, 1981 for summary). Although well-known, the four mangrove genera (*Rhizophora*, *Bruguiera*, *Kandelia*, and *Ceriops*; about 17 species) actually compose only a portion of the family, with the majority of species being shrubs and climbers of (inland) tropical rain forests (Graham, 1964). The term "mangrove" has also been applied to several genera of other families, notably *Avicennia* (Verbenaceae) and *Laguncularia* and *Conocarpus* (both in the Combretaceae).

Rhizophora trees and shrubs exhibit a characteristic "mangrove habit" (Gill and Tomlinson, 1969, 1972). A plant is supported by large downward curving prop roots (or aerial roots) that arise from the main trunk or from large branches. These roots eventually interlace around the base of the trunk, forming a series of emergent loops. Conspicuous white lenticels on the aerial roots function in gas exchange to the absorptive underground root system.

The flowers of *Rhizophora mangle* are anemophilous, although visiting bees probably collect the pollen (Tomlinson et al., 1979; Tomlinson, 1980). Each anther possesses numerous cross partitions and dehisces with a flap before anthesis. The hairy inner surface of the corolla traps the pollen grains, which are dispersed as the petals recurve. The stigma becomes receptive after the pollen is shed, promoting cross-pollination.

The seeds of *Rhizophora* germinate while on the tree (vivipary) without undergoing a dormant period. As the embryo develops, the endosperm grows out of the micropyle and forms an aril-like covering on the seed. The cotyledons (fused into a tube) and then the hypocotyl emerge through the micropyle. When the hypocotyl is fully elongate, the seedling (hypocotyl plus plumule) detaches from the cotyledons (and the rest of the fruit) and falls from the tree into the water. Thus, the seedling itself is the unit of dispersal (Rabinowitz, 1978).

References Cited

Cronquist, A. 1981. An integrated system of classification of flowering plants. Pp. 655-659. Columbia Univ. Press, New York, NY.

Gill, A.M. and P.B. Tomlinson. 1969. Studies on the growth of red mangrove (*Rhizophora mangle* L.). 1. Habit and general morphology. Biotropica 1: 1-9.

_____ and _____. 1972. Ibid. 2. Growth and differentiation of aerial roots. Biotropica 3: 63-77.

Graham, S.A. 1964. The genera of Rhizophoraceae and Combretaceae in the southeastern United States. J. Arnold Arbor. 45: 285-301.

Rabinowitz, D. 1978. Dispersal properties of mangrove propagules. Biotropica 10: 47-57.

Tomlinson, P.B., R.B. Primack, and J.S. Bunt. 1979. Preliminary observations on floral biology in mangrove Rhizophoraceae. Biotropica 11: 256-277.

Tomlinson, P.B. 1980. The biology of trees native to tropical Florida. Pp. 320-326. "Publ. by the author", Petersham, MA.

VITACEAE (GRAPE FAMILY)

Climbing woody vines, usually with tendrils opposite the leaves; stems sympodial, with swollen or jointed nodes. <u>Leaves</u> simple or pinnately or palmately compound, often palmately lobed and/or veined, generally alternate (lower ones sometimes opposite), distichous, pellucid punctate, stipulate (stipules small, caducous). <u>Inflorescence</u> determinate, cymose and sometimes appearing racemose or paniculate, opposite a leaf or sometimes also terminal. <u>Flowers</u> actinomorphic, perfect or imperfect (then plants usually monoecious or polygamodioecious), hypogynous (or slightly perigynous), minute, with well-developed intrastaminal disc or sometimes 5 distinct glands. <u>Calyx</u> of 4 or 5 often indistinct teeth or lobes, sometimes reduced to a rim around the ovary, cup-like. <u>Corolla</u> typically of 4 or 5 petals, distinct or sometimes apically connate and caducous as a calyptra (cap), usually greenish, valvate. <u>Androecium</u> of 4 or 5 stamens, opposite the petals; filaments distinct; anthers dorsifixed, distinct or connate, dehiscing longitudinally, introrse; staminodes present in carpellate flowers. <u>Gynoecium</u> of 1 pistil, 2-carpellate; ovary superior, 2-loculed (septa often incomplete), variously adnate to the disc; ovules 1 or 2 on each placenta, anatropous, erect, placentation axile; style 1, short; stigma 1, small, discoid or capitate; rudimentary pistil present in staminate flowers. <u>Fruit</u> a berry; seeds 1 to 4, with hard, bony, or crustaceous testa, with conspicuous "chalazal knot" and 2 deep adaxial grooves; endosperm copious, usually 3-lobed, hard-fleshy, oily, proteinaceous; embryo small, straight, spatulate, surrounded by endosperm.

Family characterization in Florida: sympodial woody vines with tendrils and swollen nodes; pellucid punctate, simple to compound, often palmately veined or lobed leaves; inflorescences arising opposite the leaves; small flowers with 4- or 5-merous perianth and 4 or 5 stamens opposite the petals; berry with seeds having characteristic morphology (see discussion below). Parenchymatous tissues with tanniferous cells, mucilage cells and raphide-sacs.

75

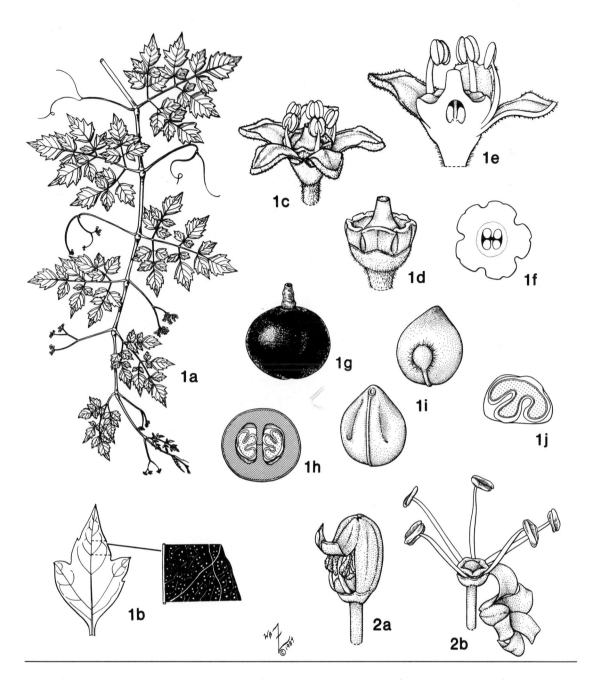

Figure 30. Vitaceae. 1, *Ampelopsis arborea:* **a,** habit, x 1/3; **b,** leaflet, x 1 1/2, with detail showing punctate surface, x 6; **c,** flower, x 9; **d,** pistil and nectary, x 12; **e,** longitudinal section of flower, x 12; **f,** cross section of ovary, x 12; **g,** berry, x 2; **h,** cross section of berry, x 2; **i,** two views of seed, x 4 1/2; **j,** cross section of seed, x 4 1/2. **2,** *Vitis rotundifolia:* **a,** staminate flower bud immediately before anthesis (corolla starting to detach), x 12; **b,** staminate flower with corolla detached as a calyptra, x 12.

Genera/species: 13/735

Distribution: Primarily pantropical with a few representatives in temperate regions.

Major genus: *Cissus* (350 spp.)

Florida representatives: *Vitis* (7 spp.), *Cissus* (3 spp.), *Ampelopsis* (2 spp.), and *Parthenocissus* (1 sp.)

Economic plants and products: Wine, juice, and table grapes, and muscatel, raisins, and "currants" from fruits of several *Vitis* spp. Ornamental vines (species of 7 genera), including: *Cissus* (grape-ivy), *Parthenocissus* (Boston-ivy, Virginia-creeper, woodbine), and *Vitis*.

The Vitaceae compose a natural family. The genera are separated primarily on the basis of the morphology of the style (relative length), nectariferous disc, and endosperm (lobing), as well as characters of the corolla (free or apically connate), inflorescence, and tendrils (attach by coiling or with disc-like structures).

Most of the Vitaceae are vines with tendrils (and inflorescences) that occur on the stem opposite the leaves, an unusual situation that suggests that the stem is sympodial (Brizicky, 1965). The tendril has been generally regarded as representing a main axis that has been subordinated (or pushed aside) by the more vigorous growth of the branch in the opposing leaf axil. However, the situation in *Cissus* and *Vitis* is more complex with the leaves having axillary buds as well as opposing tendrils (Cronquist, 1981). Since flower-bearing tendrils may occur, it has also been suggested that the tendrils represent modified inflorescences.

The flowers, which secrete nectar from the disc and sometimes have a sweet odor (*Vitis*), are generally entomophilous. However, a few species are apparently anemophilous.

The seeds of the Vitaceae have a distinctive morphology. The seed coat consists of a thin outer membrane and a hard (often bony) inner layer (sclerotesta) that forms two deep grooves on the adaxial surface. This infolding conforms the endosperm into a characteristic three-lobed configuration. A conspicuous ridge (raphe) occurs between the grooves and extends from the hilum to the seed apex (on the adaxial side). It terminates on the abaxial surface in a conspicuous, depressed or elevated area called a "chalazal knot" (see Fig. 30, 1i and 1j).

References Cited

Brizicky, G.K. 1965. The genera of Vitaceae in the southeastern United States. J. Arnold Arbor. 46: 48-67.

Cronquist, A. 1981. An integrated system of classification of flowering plants. Pp. 748-750. Columbia Univ. Press, New York, NY.

CORNACEAE (DOGWOOD FAMILY)

Usually trees or shrubs or sometimes suffrutescent sub-shrubs. <u>Leaves</u> simple, entire to obscurely serrate, usually opposite, often with arching secondary veins, usually deciduous, exstipulate. <u>Inflorescence</u> determinate, basically cymose and appearing corymbose, umbellate, capitate, or paniculate, sometimes subtended by large (usually white) petaloid bracts, terminal or axillary. <u>Flowers</u> actinomorphic, perfect or sometimes imperfect (then plants monoecious, dioecious or polygamodioecious), epigynous, small, with a nectariferous disc at ovary apex, sessile or nearly so. <u>Calyx</u> of usually 4 sepals, distinct, small and inconspicuous, often reduced to teeth around upper edge of ovary, valvate. <u>Corolla</u> of usually 4 petals, distinct, white, purple, or yellow, valvate. <u>Androecium</u> of 4 stamens; filaments distinct; anthers usually dorsifixed, versatile, dehiscing longitudinally. <u>Gynoecium</u> of 1 pistil, usually 2-carpellate; ovary inferior,

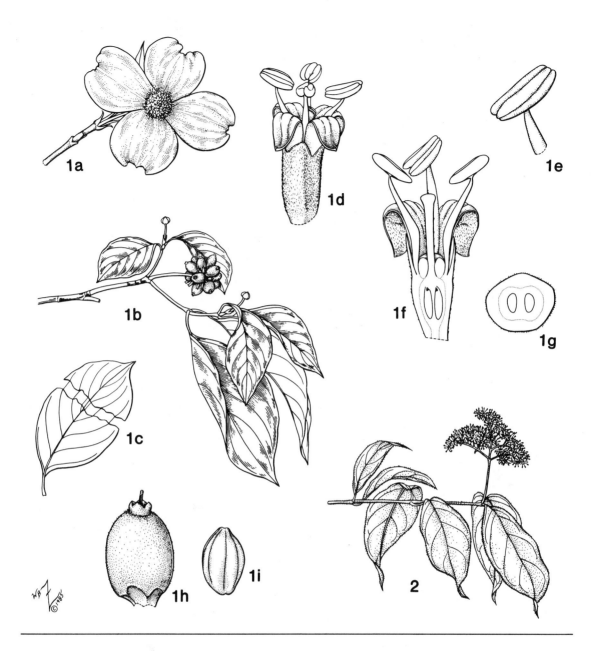

Figure 31. Cornaceae. 1, *Cornus florida:* **a,** flowering branch, x 1/2; **b,** fruiting branch, x 1/3; **c,** torn leaf connected by unraveled spiral thickenings of the vessel elements ("*Cornus* test"), x 1/3; **d,** flower, x 6; **e,** anther, x 12; **f,** longitudinal section of flower, x 8; **g,** cross section of ovary, x 10; **h,** drupe, x 2 1/2; **i,** pyrene, x 2 1/2. **2,** *Cornus foemina:* flowering branch, x 1/2.

usually 2-loculed, capped by nectariferous disc at apex; ovules solitary in each locule, anatropous, pendulous, placentation axile; style 1; stigmas capitate or lobed. <u>Fruit</u> usually a drupe with 1 furrowed pyrene (stone); endosperm copious, oily; embryo small, spatulate.

Family characterization in Florida: trees or shrubs; usually opposite simple leaves with arching (arcuate) secondary veins; 4-merous perianth and androecium; versatile anthers; 2-carpellate inferior ovary capped by a nectariferous disc; and a drupe with a single furrowed stone. Anatomical features: unitegmic ovules and trilacunar nodes.

Genera/species: 4/72

Distribution: Widespread; most common in north temperate regions.

Major genus: *Cornus* (45 spp.)

Florida representatives: *Cornus* (4 spp.)

Economic plants and products: Edible fruits from *Cornus mas* (cornelian-cherry). Wood from several *Cornus* spp. Ornamental trees and shrubs (species of 2 genera): *Cornus* (dogwood) and *Curtisia.*

In this treatment, the Cornaceae are presented as delimited by Thorne (1983; see also Ferguson, 1977). The Cornaceae and *Cornus* have been variously divided in the literature (see Ferguson, 1966a,b and Rodriguez, 1971). The eight sections (or subgenera) of *Cornus s.l.* have also been recognized as segregate genera by some authors; many genera (including these *Cornus* segregates) of Cornaceae *s.l.* have now been separated into small or monotypic families. The very closely related Nyssaceae (with 3 species of *Nyssa* in Florida), for example, were once generally included within the Cornaceae.

Sterile *Cornus* plants are readily recognizable by the arcuate venation of the leaves, and the "*Cornus* test" (see Fig. 31, 1c) may verify the identification of the genus. If the leaf blade is carefully torn in two, the main veins remain connected by delicate threads (unraveled spiral thickenings in the vessel elements).

The flowers of *Cornus* are generally homogamous (anthers and stigmas mature simultaneously), and attract various insects (bees, flies, and beetles) with the nectar secreted by the disc surrounding the style. In many species, the flowers are also very fragrant. The capitate inflorescences of several *Cornus* species, such as *C. florida*, are encircled by large, petaloid, usually white bracts. These form a conspicuous pseudanthium, with the bracts functioning as "petals" and the small flowers in the center, as "stamens and pistils." Cross- and self-pollination may occur as the insects walk over the surface of the inflorescence, although cross-pollination is somewhat favored by the different filament and style lengths.

References Cited

Ferguson, I.K. 1966a. Notes on the nomenclature of *Cornus*. J. Arnold Arbor. 47: 100-105.

————. 1966b. The Cornaceae in the southeastern United States. J. Arnold Arbor. 47: 106-116.

————. 1977. Cornaceae. World Pollen and Spore Flora 6: 1-34.

Rodriguez, R.L. 1971. The relationships of the Umbellales. Pp. 63-91 *in:* V.H. Heywood (ed.), The biology and chemistry of the Umbelliferae. Academic Press, Inc., London, England.

Thorne, R.F. 1983. Proposed new realignments in the angiosperms. Nord. J. Bot. 3: 85-117.

IRIDACEAE (IRIS FAMILY)

Perennial herbs, often scapose, with rhizomes, bulbs, or corms; roots fibrous, mycorrhizal. <u>Leaves</u> simple, entire, alternate and basal, distichous, equitant, parallel veined, numerous, usually narrow, conduplicate, with open basal sheath, exstipulate. <u>Inflorescence</u> determinate, cymose and often appearing racemose or paniculate, or sometimes flowers solitary, terminal. <u>Flowers</u> actinomorphic or zygomorphic, perfect, epigynous, usually large and showy, individually or collectively subtended by 1 or 2 expanded bract(s) (forming a spathe). <u>Perianth</u> of 6 tepals in 2 whorls of 3, similar or differentiated, generally all petaloid, distinct or usually basally connate and forming a tube, variously colored, overlapping and twisted in bud. <u>Androecium</u> of 3 stamens, opposite and often adnate to the outer tepals; filaments distinct or sometimes basally connate and forming a tube; anthers basifixed, dehiscing longitudinally, extrorse. <u>Gynoecium</u> of 1 pistil, 3-carpellate; ovary inferior, 3-loculed; ovules few to many on each placenta, anatropous, placentation axile; style 1, often trifid with branches simple, flattened, or enlarged and petaloid; stigmas 3, terminal (style branches simple) or on abaxial surface of petaloid style branches, papillate. <u>Fruit</u> a 3-valved loculicidal capsule; seeds numerous, sometimes arillate or with sarcotesta; endosperm copious, fleshy; embryo small, linear, straight.

Family characterization in Florida: perennial herbs with rhizomes, bulbs, or corms; equitant, basal, linear to ensiform leaves; 3-merous flowers with petaloid perianth and 3 stamens; inferior ovary; and a loculicidal capsule as the fruit type. Tissues commonly with tannins and calcium oxalate crystals (long and prismatic).

Genera/species: 81/1,500

Distribution: Cosmopolitan; centers of diversity in South Africa, the eastern Mediterranean region, and tropical America.

Major genera: *Iris* (200-300 spp.), *Gladiolus* (150-300 spp.), *Moraea* (100 spp.), and *Sisyrinchium* (60-100 spp.)

Florida representatives: 9 genera/22 spp.; largest genera: *Sisyrinchium* (8 spp.) and *Iris* (7 spp.)

Economic plants and products: Saffron (a dye and a spice) from the stigmas of *Crocus sativus*. Orris root (a fragrant substance used in perfumes and cosmetics) from rhizomes of *Iris* spp. Ornamental plants (species of 46 genera), including: *Crocus*, *Dietes*, *Eustylis*, *Freesia*, *Gladiolus*, *Iris* (flag, fleur-de-lis), *Ixia* (corn-lily), *Moraea* (butterfly-iris, natal-lily), *Neomarica* (fan-iris), *Sisyrinchium* (blue-eyed grass), and *Tigridia* (tiger-flower).

Botanists generally concur that the Iridaceae are related to and more advanced than the Liliaceae (Cronquist, 1981; Dahlgren and Clifford, 1982; Dahlgren et al., 1985). The family is divided into subfamilies, tribes, and genera based on the rootstock (rhizomes, bulbs, or corms), inflorescence type, floral symmetry, and style morphology (e.g., three-branched with terminal stigmas or petaloid with sub-terminal stigmas).

The brightly colored flowers are usually pollinated by insects (such as bees and flies) that seek the nectar accumulated at the base of the floral tube. Nectaries may occur in the form of septal glands (*Gladiolus*) or at the base of the tepals or stamens (*Iris*). The pollination mechanism depends upon the particular floral morphology. For example, the perianth of *Sisyrinchium* consists of six similar tepals forming a short tube with spreading lobes. Often a yellow or white "eye" targets the center of the flower.

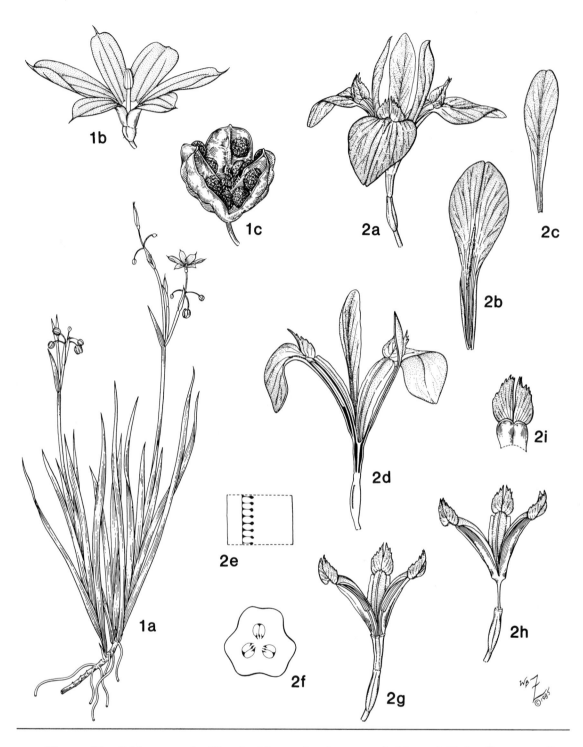

Figure 32. Iridaceae. 1, *Sisyrinchium atlanticum:* **a,** habit, x 1/2; **b,** flower, x 3; **c,** capsule, x 4 1/2. **2,** *Iris hexagona* var. *savannarum:* **a,** flower, x 1/2; **b,** outer tepal, x 1/2; **c,** inner tepal, x 1/2; **d,** longitudinal section of flower, x 2/3; **e,** detail of longitudinal section of ovary, x 3; **f,** cross section of ovary, x 3; **g,** androecium and gynoecium, x 1/2; **h,** pistil, x 1/2; **i,** underside of style branch apex showing receptive area (flap), x 3/4.

The monadelphous stamens surround the style, which has filiform branches. The extrorse anthers typically release the pollen before the style branches expand.

In comparison, a highly modified perianth and style characterize the complex *Iris* flower. The large tepals (called the "falls") of the outer whorl are spreading or deflexed, while those of the inner whorl (the "standards") are smaller and erect. Each outer tepal is often bearded with a crest of hairs and marked with nectar guides (lines). The flat and petaloid style branches curve along the outer tepals, forming a protective covering over the stamens. A flap-like stigma is situated on the underside of each style branch near the distal end. Only the adaxial surfaces of the stigmas are receptive.

An insect uses a large tepal of an *Iris* flower as a landing platform. It may then transfer pollen from a previous flower to the receptive stigmatic surface while it probes for the nectar through the tube (or passageway) formed by the tepal and the corresponding style branch. Further down, it becomes dusted with new pollen from the anther. As the insect retreats from the flower, it encounters only the non-receptive abaxial surface of the stigma.

References Cited

Cronquist, A. 1981. An integrated system of classification of flowering plants. Pp. 1211-1213. Columbia Univ. Press, New York, NY.

Dahlgren, R.M.T. and H.T. Clifford. 1982. The monocotyledons: a comparative study. Pp. 26-30. Academic Press, Inc., London, England.

_____, _____, and P.F. Yeo. 1985. The families of the monocotyledons. Pp. 238-249. Springer-Verlag, Berlin, Germany.

HYDROCHARITACEAE (FROG'S-BIT OR TAPE-GRASS FAMILY)

Perennial aquatic herbs, completely or partially submersed or floating, with creeping monopodial rhizome and/or erect main stem. Leaves simple, entire to serrate, crowded and basal or cauline (then alternate, opposite, or whorled), generally parallel veined, extremely variable in shape and size, with sheathing bases. Inflorescence determinate, cymose and umbellate (staminate inflorescence) or flower solitary (carpellate inflorescence), subtended by 2 distinct or connate bracts (forming a spathe), axillary. Flowers generally actinomorphic, usually imperfect (then plants most often dioecious), epigynous, small and inconspicuous to fairly large and showy, sometimes with very long peduncles. Calyx of 3 sepals, distinct, valvate. Corolla of 3 petals or sometimes absent, distinct, delicate and fugacious, usually white, imbricate or convolute. Androecium of 2 or 3 to numerous stamens, often in 1 to 5 whorls of 3, inner or outer whorls sometimes reduced to staminodes; filaments distinct to sometimes connate (monadelphous); anthers dehiscing longitudinally; staminodes often present in carpellate flowers. Gynoecium of 1 pistil, usually 3- to 6-carpellate; ovary inferior, 1-loculed; ovules numerous, anatropous, placentation parietal and often appearing axile due to 3 to many deeply intruding placentae, styles as many as carpels, often bifid; stigmas papillate; rudimentary pistil sometimes present in staminate flowers. Fruit berry-like, dry or fleshy, indehiscent or rupturing irregularly, globose to linear, submersed; seeds usually numerous; endosperm absent; embryo straight.

Family characterization in Florida: submersed to floating aquatic herbs; inflorescence subtended by a spathe or a pair of bracts; imperfect flowers; syncarpous gynoecium with inferior, 1-loculed ovary; parietal placentation with placentae deeply intruded into the ovary; and berry-like submersed fruit. Anatomical features: scattered tanniferous cells and aerenchymatous stem with much reduced vascular system.

82

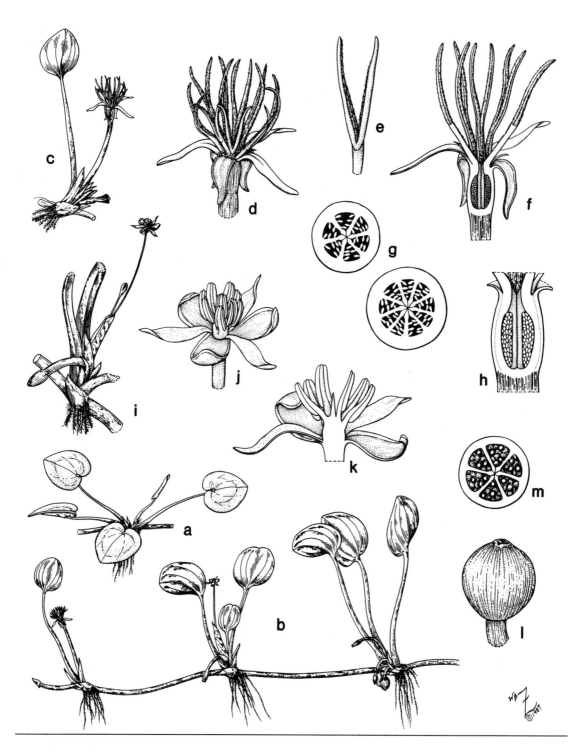

Figure 33. Hydrocharitaceae. *Limnobium spongia:* **a,** juvenile plant, x 1/4; **b,** habit with flowers and fruit, x 1/4; **c,** node with cluster of carpellate inflorescences, x 2/3; **d,** carpellate flower, x 2; **e,** stigma, x 3; **f,** longitudinal section of carpellate flower, x 2 1/2; **g,** cross section of ovaries from two different flowers, x 6; **h,** longitudinal section of ovary, x 4 1/2; **i,** node with staminate inflorescence and carpellate (fruiting) inflorescence, x 2/3; **j,** staminate flower, x 3; **k,** longitudinal section of staminate flower, x 4; **l,** fruit, x 1 1/2; **m,** cross section of fruit, x 1 1/2.

Genera/species: 16/80

Distribution: Primarily in tropical marine and freshwater habitats with a few representatives in temperate freshwaters.

Major genus: *Otellia* (40 spp.)

Florida representatives: *Halophila* (2 or 3 spp.), *Vallisneria* (1 or 2 spp.), *Elodea* (1 sp.), *Egeria* (1 sp.), *Hydrilla* (1 sp.), *Limnobium* (1 sp.), and *Thalassia* (1 sp.)

Economic plants and products: Several weedy plants, such as species of *Elodea* (waterweed, ditch-moss) and *Hydrilla*. Ornamental plants for aquaria (species of 9 genera), including: *Elodea*, *Egeria* (waterweed), *Hydrilla*, *Limnobium* (American frog's-bit), and *Vallisneria* (eel-grass, tape-grass).

The Hydrocharitaceae are divided into three to five subfamilies (see Dahlgren et al., 1985). The two monotypic groups (Thalassioideae and Halophiliodeae) are characterized by growing in salt water.

The floating to submersed vegetative habit varies considerably in the family (Ancibor, 1979). For example, the linear and grass-like leaves of *Vallisneria* are clustered on short stems along the nodes of the rhizome. The leaves are also clustered in the marine genera, *Thalassia* (with ribbon-like leaves) and *Halophila* (leaves clustered at summits of stems). In *Hydrilla* or *Elodea*, the small sessile leaves are in whorls along the submersed and free-floating stems. The leaves of *Limnobium* are clearly differentiated into a lamina and petiole. The juvenile form consists of floating rosettes of reniform leaves, each with a central area of spongy tissue. Eventually, more robust plants develop that bear the flowers and fruit. The leaves of these mature plants are long-petiolate with rounded and leathery blades.

The flowers vary considerably in carpel and stamen number and are typically imperfect (Kaul, 1968, 1970). Generally, the inflorescences are unisexual, with each subtended by a spathe. The carpellate inflorescences are usually one-flowered, and the staminate, one- to many-flowered.

The Hydrocharitaceae are notable for specialized pollination mechanisms (Sculthorpe, 1967). Several with relatively large and showy flowers (e.g., *Egeria* and most *Limnobium* spp.) have nectaries and are pollinated by insects (such as flies and beetles). In others, such as *Hydrilla* and *Vallisneria*, the staminate flowers become detached and drift in the water to the carpellate inflorescences (Cook and Luond, 1982; Lowden, 1982). The long peduncles of the carpellate flowers of *Vallisneria* coil up after pollination, pulling the flowers beneath the surface of the water. The anthers of *Elodea* (staminate flowers sometimes detach) and *Hydrilla* explode and scatter pollen over the surface of the water. Pollination takes place underwater in the flowers of both *Thalassia* and *Halophila*, which release the pollen in connected chains (Tomlinson, 1969).

References Cited

Ancibor, E. 1979. Systematic anatomy of vegetative organs of the Hydrocharitaceae. J. Linn. Soc., Bot. 78: 237-266.

Cook, C.D.K. and Luond, R. 1982. A revision of the genus *Hydrilla* (Hydrocharitaceae). Aquatic Bot. 13: 485-504.

Dahlgren, R.M.T., H.T. Clifford, and P.F. Yeo. 1985. The families of the monocotyledons. Pp. 303-307. Springer-Verlag, Berlin, Germany.

Kaul, R.B. 1968. Floral morphology and phylogeny in the Hydrocharitaceae. Phytomorphology 18: 13-35.

———. 1970. Evolution and adaptation of the inflorescences in the Hydrocharitaceae. Amer. J. Bot. 57: 708-715.

Lowden, R.M. 1982. An approach to the taxonomy of *Vallisneria* L. (Hydrocharitaceae). Aquatic Bot. 13: 269-298.

Sculthorpe, C.D. 1967. The biology of aquatic vascular plants. Pp. 301-311. St. Martin's Press, New York, NY.

Tomlinson, P.B. 1969. On the morphology and anatomy of turtle grass *Thalassia testudinum* (Hydrocharitaceae). III. Floral morphology and anatomy. Bull. Mar. Sci. Gulf Caribbean 19: 286-305.

LEMNACEAE (DUCKWEED FAMILY)

Perennial aquatic herbs, free-floating or submersed, undifferentiated into stem and leaves and reduced to a thallus ("frond"), small to minute, flat and leaf-like to globose, sometimes purplish beneath, often reproducing asexually (by budding); roots simple and thread-like or absent. Flowers only occasionally produced, solitary or paired in pouches (on margins and/or upper surface of thallus), imperfect (plants monoecious), hypogynous, often initially enclosed by a membranous sheath (spathe). Perianth absent. Androecium of usually 1 stamen; filaments filiform to fusiform or absent; anthers 2- or sometimes 1-loculed, dehiscing longitudinally or transversely. Gynoecium of 1 pistil, 1-carpellate; ovary superior, 1-loculed; ovule(s) usually 1 or 2, orthotropous, placentation basal; style 1, short; stigma 1, funnel-shaped. Fruit a utricle; seed(s) usually 1 or 2, with cap-like operculum at micropylar end ("stopper" or inner integument); endosperm scanty, fleshy, sheathing the embryo, or absent; embryo relatively large, straight, consisting almost entirely of a large cotyledon.

Family characterization in Florida: submersed or free-floating aquatic herbs occurring in still freshwaters; reduced plant body (without stem or leaves) of a small leaf-like thallus; frequent asexual reproduction (by budding) and only occasional flowering; imperfect flowers consisting of only of one stamen (staminate flowers) or of one carpel (carpellate flowers); and a utricle as the fruit type. Anatomical features: aerenchyma tissue (in thallus) and typically a reduced vascular system without xylem (although tracheids occur in the roots of *Spirodela*).

Genera/species: 6/28

Distribution: Cosmopolitan; in quiet freshwater habitats.

Major genera: *Lemna* (9 spp.), *Wolffia* (7 spp.), and *Wolffiella* (5 spp.)

Florida representatives: *Lemna* (7 spp.), *Wolffia* (3 spp.), *Wolffiella* (2 or 3 spp.), and *Spirodela* (2 spp.)

Economic plants and products: Important food sources for waterfowl and fish, serious weeds of still waters, and ornamental plants for pools and aquaria (species of all 6 genera), including: *Lemna* (duckweed, duck's-meat), *Spirodela* (duckweed), *Wolffia* (water-meal), and *Wolffiella* (mud-midget, bogmat).

The Lemnaceae are the smallest of seed plants, with *Wolffia* species being the tiniest of all. They are closely related to and derived from the Araceae. Plants of *Spirodela*, the least reduced genus, are somewhat similar to those of *Pistia*, the

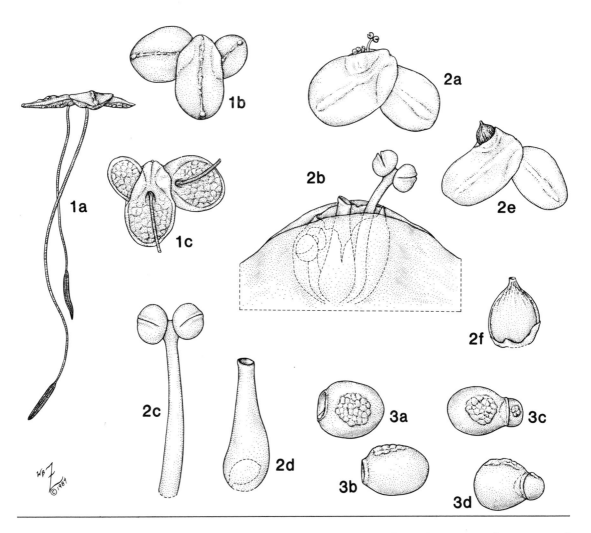

Figure 34. Lemnaceae. 1, *Lemna obscura:* **a,** lateral view of a clone of three thalli, x 8; **b,** dorsal view, x 8; **c,** ventral view, x 8. **2,** *Lemna minor* (from FLAS 71765): **a,** dorsal view of flowering thallus, x 12; **b,** detail of margin of thallus showing pouch with inflorescence, x 45; **c,** staminate flower, x 60; **d,** carpellate flower, x 60; **e,** dorsal view of fruiting thallus, x 12; **f,** utricle, x 25. **3,** *Wolffia columbiana:* **a,** dorsal view of thallus, x 22; **b,** lateral view of thallus, x 22; **c,** dorsal view of thallus with a vegetative bud, x 22; **d,** lateral view of thallus with a vegetative bud, x 22.

monotypic genus of free-floating aroids (see Wilson, 1960; Hartog, 1975; Dahlgren et al., 1985; and Fig. 36, 3 in *Part I*).

Duckweed plants form dense floating mats on ponds or pools. The vegetative structure consists of a specialized plant body (frond or thallus) of various shapes (circular, elongate, or ellipsoid) that represents a modified stem and/or leaf (see Hartog and Plas, 1970 and Landolt, 1980). Vegetative buds (and in a few genera, flowers) develop in one or two reproductive pouches along the basal margins of the thallus. Most species have no roots, although the thalli of *Spirodela* have several, and those of *Lemna*, a single "root" with no vascular tissue (Hillman, 1961).

Reproduction is primarily vegetative. New thalli, which emerge from the reproductive pockets, may separate or remain attached to the parent plant. Bulblets, resting thalli with abundant starch reserves, are produced before winter or during conditions unfavorable for growth. Depending on the species, the bulblets either sink to the bottom of the pond or remain in sheltered areas until favorable conditions return.

The flowers of the Lemnaceae, which develop in pockets on the thallus, are rarely produced (Saeger, 1929; Hicks, 1932). The staminate flowers consist of a single stamen, and the carpellate, of a single carpel. The inflorescence of *Lemna* or *Spirodela* consists of two staminate flowers and one carpellate flower subtended by a minute, rudimentary spathe. The spathe is absent in *Wolffia* and *Wolffiella*, which have only one staminate and one carpellate flower in each inflorescence. Flowers in the same plant mature at different times. Insects and other aquatic animals, as well as direct contact of the flowers, may cause cross-pollination.

The seeds are released from the fruit by the disintegration of the pericarp.

References Cited

Dahlgren, R.M.T., H.T. Clifford, and P.F. Yeo. 1985. The families of the monocotyledons. Pp. 287-289. Springer-Verlag, Berlin, Germany.

Hartog, C. den. 1975. Thoughts about the taxonomical relationships within the Lemnaceae. Aquatic Bot. 1: 407-416.

_____ and F. van der Plas. 1970. A synopsis of the Lemnaceae. Blumea 18: 355-368.

Hicks, L.E. 1932. Flower production in the Lemnaceae. Ohio J. Sci. 32: 115-131.

Hillman, W.S. 1961. The Lemnaceae or duckweeds. A review of the descriptive and experimental literature. Bot. Rev. 27: 221-287.

Landolt, E. 1980. Key to the determination of taxa within the family of Lemnaceae. Veröff. Geobot. Inst. ETH Stiftung Rübel Zürich 70: 13-21.

Saeger, A. 1929. The flowering of Lemnaceae. Bull. Torrey Bot. Club 56: 351-358.

Wilson, K.A. 1960. The genera of Arales in the southeastern United States. J. Arnold Arbor. 41: 47-72.

BROMELIACEAE (BROMELIAD OR PINEAPPLE FAMILY)

Perennial herbs, epiphytic or sometimes terrestrial xerophytes, scapose; stems usually reduced or absent; roots adventitious. Leaves simple, entire or spinose-serrate, alternate and spirally arranged, clustered and forming basal rosettes, parallel veined, sessile, strap-shaped and often trough-like, stiff, often reddish or purplish at base, with basal sheaths often overlapping and forming a cup. Inflorescence indeterminate, spicate, racemose, paniculate or capitate, usually terminal. Flowers actinomorphic or slightly zygomorphic, usually perfect, hypogynous to more often epigynous, each usually in the axil of a brightly colored bract. Calyx of 3 sepals, distinct or basally connate, convolute and/or imbricate, persistent. Corolla of 3 sepals, distinct or basally connate, frequently with a pair of scale-like appendages at base, often brightly colored, convolute and/or imbricate. Androecium of 6 stamens, biseriate, often epipetalous; filaments distinct or basally connate; anthers basifixed or dorsifixed, versatile, dehiscing longitudinally, introrse. Gynoecium of 1 pistil, 3-carpellate; ovary inferior to sometimes superior, 3-loculed; ovules usually numerous, anatropous, placentation axile; style 1; stigmas 3, often spirally twisted (at least in bud). Fruit a septicidal capsule or a berry (sometimes becoming a multiple fruit by connation of berries and axes of adjacent flowers); seeds often winged, caudate, or with plumose appendage; endosperm copious, mealy; embryo small, usually peripheral at base of endosperm.

87

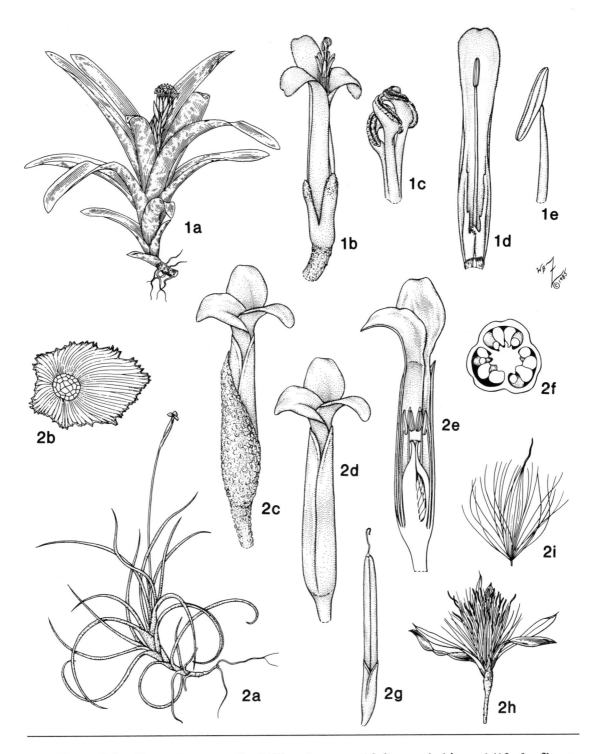

Figure 35. Bromeliaceae. 1, *Billbergia pyramidalis:* **a,** habit, x 1/10; **b,** flower, x 1 1/4; **c,** apex of style showing the stigmas, x 6; **d,** petal with adnate stamens and paired appendages, x 1 1/2; **e,** anther, x 5. **2,** *Tillandsia recurvata:* **a,** habit, x 2/3; **b,** top view of peltate scale from leaf, x 45; **c,** flower with subtending bract, x 6; **d,** flower, x 6; **e,** longitudinal section of flower, x 7; **f,** cross section of ovary, x 18; **g,** capsule, x 2; **h,** dehisced capsule, x 1 1/4; **i,** seed, x 1 1/2.

Family characterization in Florida: epiphytic scapose herbs with reduced stems and adventitious roots; strap-like, stiff, usually concave leaves often with colored and sheathing bases; 3-merous flowers subtended by bracts; and seeds with plumose appendages. Anatomical features: mucilage cells and raphide-sacs (in all organs), silica bodies (in epidermal cells), and tannin granules (in parenchymatous cells of shoot), and vestiture of peltate scales (as well as numerous other adaptations of the leaves for water retention and absorption; see discussion below).

Genera/species: 44/2,000

Distribution: Almost exclusively (except for one species of *Pitcairnia*) in tropical to warm temperate America.

Major genera: *Tillandsia* (500 spp.), *Pitcairnia* (250 spp.), *Vriesea* (190 spp.), and *Aechmea* (150 spp.)

Florida representatives: *Tillandsia* (13 spp.), *Catopsis* (3 spp.), *Guzmania* (1 sp.), *Ananas* (1 sp.), *Billbergia* (1 sp.), and *Bromelia* (1 sp.)

Economic plants and products: Edible fruits from *Ananas comosa* (pineapple). Cordage and fiber for fabric from several, such as species of *Ananas* and *Aechmea*. "Vegetable hair" (dried stems and leaves) used for upholstery stuffing from *Tillandsia* (Spanish-moss). Ornamental plants (species of 33 genera), including: *Aechmea* (air-pine, living-vase), *Billbergia* (vase-plant), *Bromelia* (pinquin), *Cryptanthus* (earth-star), *Guzmania*, *Nidularium*, *Pitcairnia*, *Tillandsia*, and *Vriesea*.

The Bromeliaceae, a very natural and distinctive family, are usually divided into three subfamilies (see Dahlgren et al., 1985) based primarily upon features of the habit (terrestrial or epiphytic), ovary (superior to inferior), fruit (capsule or berry), and seeds (type of appendages).

Bromeliads are xerophytes and the majority are epiphytic. The typical "tank-type" plant has a shortened axis tightly clasped by the well-developed leaf sheaths, thus forming a basin that collects water as well as decayed material. In most species, the leaf is the primary organ for absorbing water. The primary root is usually short-lived; the adventitious roots (developing at the leaf bases) function mainly to attach the plant to the substrate. Water and often dissolved organic compounds are absorbed by specialized peltate scales that occur within the leaf sheath, or in the most advanced species (e.g., *Tillandsia* spp.), over the entire surface of the plant. Each trichome generally consists of a uniseriate stalk plus an expanded shield-like apex, which is composed of dead cells at maturity. When moistened, the empty "shield cells" expand, and water is drawn osmotically through the living stalk cells to the mesophyll (Tomlinson, 1969; Benzing, 1980). The leaves typically have water-storing parenchyma (water-storage tissue) between the assimilating mesophyll and the epidermis on the adaxial side. A thick cuticle helps reduce water loss, and in some bromeliads (*Tillandsia*), the dry and collapsed scales also provide some insulation (Smith and Wood, 1975).

The flowers are protandrous, with the stigmas spirally twisted into a head at anthesis. After the pollen is shed, the stigmas expand to expose the receptive surfaces. Insects and birds (hummingbirds) are the principal pollinators, seeking the nectar secreted by the septal nectaries (in the ovary) and sometimes by the appendages of the corolla. The flowers of many species also have an attractive and/or strong scent.

The fruit is either a capsule (*Tillandsia*) or a berry (*Bromelia*). In *Ananas* (pineapple), the whole fleshy inflorescence (axis, bracts, and berries) forms a succulent syncarp. The apical crown of leaves represents the growth of the axis beyond the inflorescence.

References Cited

Benzing, D. 1980. The biology of the bromeliads. Pp. 58-77. Mad River Press, Inc., Eureka, CA.

Dahlgren, R.M.T., H.T. Clifford and P.F. Yeo. 1985. The families of the monocotyledons. Pp. 329-333. Springer-Verlag, Berlin, Germany.

Smith, L.B. and C.E. Wood. 1975. The genera of Bromeliaceae in the southeastern United States. J. Arnold Arbor. 56: 375-397.

Tomlinson, P.B. 1969. Commelinales -- Zingiberales. Pp. 193-294 *in:* C.R. Metcalfe (ed.), Anatomy of the monocotyledons. Vol. 3. Clarendon Press, Oxford, England.

XYRIDACEAE (YELLOW-EYED-GRASS FAMILY)

Perennial or sometimes annual herbs growing in damp habitats, scapose, with short and sometimes bulbous rhizomes; roots fibrous. <u>Leaves</u> simple, entire, alternate, distichous, often equitant, basally tufted, narrow, flat, terete, or filiform, parallel veined, with an open basal sheath. <u>Inflorescence</u> indeterminate, capitate or spicate, cone-like with spirally arranged and closely imbricated bracts, terminal on a long scape. <u>Flowers</u> slightly zygomorphic, perfect, hypogynous, sessile, each in the axil of a stiff or coriaceous bract. <u>Calyx</u> of 3 sepals, distinct, unequal with the 2 lower (lateral) sepals chaffy, keeled, and persistent, and the inner (interior) sepal larger, membranous, enveloping the corolla (as a hood), and fugacious. <u>Corolla</u> of 3 petals, distinct or basally connate, clawed, ephemeral, usually yellow, marcescent. <u>Androecium</u> of 3 stamens (opposite the corolla lobes) and usually 3 staminodes (bifid, plumose, or bearded with moniliform hairs), epipetalous; filaments distinct, usually short and flattened; anthers dehiscing longitudinally, extrorse or latrorse. <u>Gynoecium</u> of 1 pistil, 3-carpellate; ovary superior, 1-loculed or basally 3-loculed; ovules few to numerous, orthotropous to anatropous, placentation parietal (1-loculed ovary) or free central/basal (in ovary partitioned at base); style 1, trifid; stigmas three, truncated. <u>Fruit</u> a 3-valved loculicidal capsule, enveloped by persistent corolla tube; seeds numerous, minute, usually apiculate and longitudinally striate; endosperm copious, mealy, starchy, proteinaceous; embryo small, lenticular or shield-shaped, apical.

Family characterization in Florida: scapose herbs often growing in damp habitats; basally tufted, narrow leaves; flowers in the axils of stiff bracts congested into cone-like heads; ephemeral yellow corolla with spreading lobes; calyx of 2 chaffy, keeled sepals and 1 membranous, hood-like sepal; androecium of 3 stamens and often 3 plumose staminodes; and 3-valved capsule enclosed by persistent corolla.

Genera/species: 4/270

Distribution: Widespread in tropical and subtropical areas, with relatively few representatives in temperate regions; especially common in the southeastern United States, tropical America, and southern Africa.

Major genus: *Xyris* (250 spp.)

Florida representatives: *Xyris* (16-19 spp.); see Kral (1960, 1966) and Godfrey and Wooten (1979).

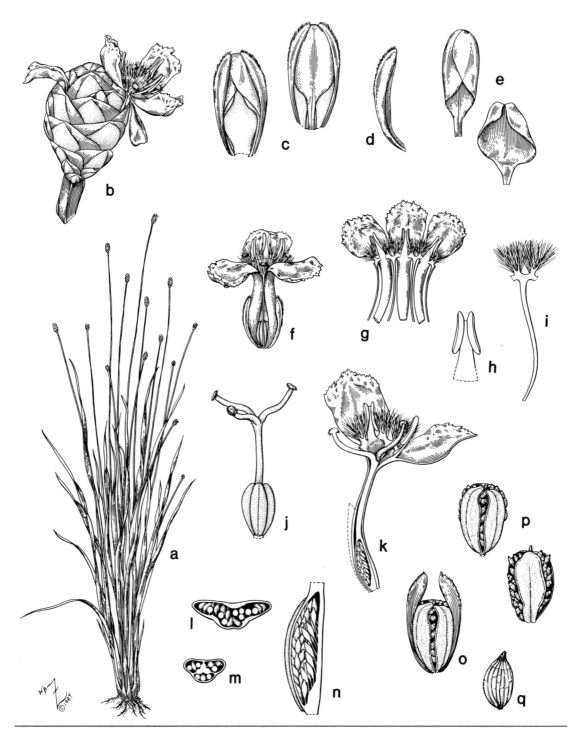

Figure 36. Xyridaceae. *Xyris platylepis:* **a,** habit, x 1/3; **b,** inflorescence, x 3; **c,** two views of bud (flower subtended by two lateral sepals and one interior sepal), x 6; **d,** one lateral sepal, x 6; **e,** interior sepal before (top) and after (bottom) anthesis, x 6; **f,** flower, x 3; **g,** expanded corolla and androecium, x 3; **h,** stamen, x 6; **i,** staminode, x 6; **j,** pistil, x 5; **k,** longitudinal section of flower, x 5; **l,** cross section of ovary near apex, x 12; **m,** cross section of ovary at base, x 12; **n,** longitudinal section of ovary, x 12; **o,** capsule with persistent lateral sepals (persistent corolla removed), x 6; **p,** two views of capsule, x 6; **q,** seed, x 30.

Economic plants and products: A few *Xyris* spp. cultivated as ornamental plants for pools and aquaria.

In the field, *Xyris* plants are unmistakable with the scapose habit and grass-like leaves, cone-like inflorescences, and yellow flowers. The corolla is ephemeral with the flower usually not open for more than a few hours. No nectar is produced. Wind-pollination appears to be prevalent, although pollen-collecting bees occasionally visit the flowers (Kral, 1983).

References Cited

Godfrey, R.K. and J.W. Wooten. 1979. Aquatic and wetland plants of the southeastern United States. Monocotyledons. Pp. 479-502. Univ. of Georgia Press, Athens, GA.

Kral, R. 1960. The genus *Xyris* in Florida. Rhodora 62: 295-319.

_____. 1966. *Xyris* (Xyridaceae) of the continental United States and Canada. Sida 2: 177-260.

_____. 1983. The Xyridaceae in the southeastern United States. J. Arnold Arbor. 64: 421-429.

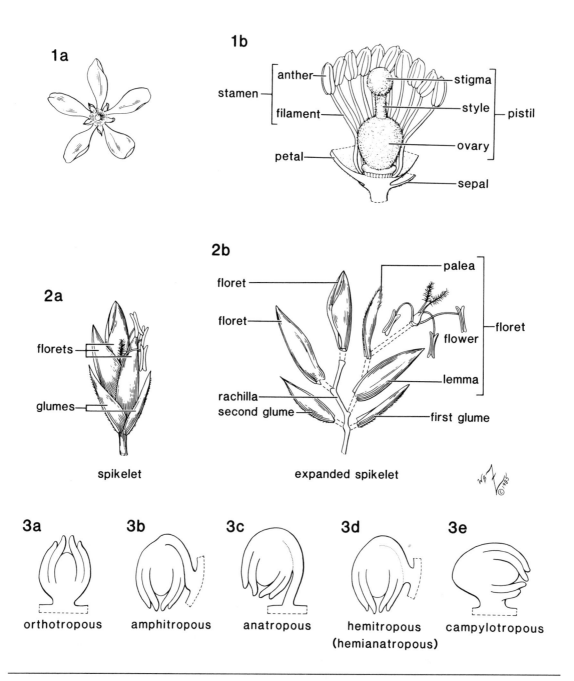

Figure 37. Additional Glossary Illustrations. 1, *Poncirus trifoliata:* **a,** flower, x 1/2; **b,** longitudinal section of flower (pistil intact), x 3. **2,** *Eragrostis spectabilis:* **a and b,** spikelet, x 12. **3a-e,** Ovule types.

GLOSSARY

These definitions apply to the angiosperms in this work, although many of these terms may also have different, more general, or more specific interpretations in other texts. Many terms relating to either special features defined in this text or basic botany (or morphology) are not repeated here. For more thorough treatments, refer to the glossaries in taxonomy texts such as Lawrence (1951), Cronquist (1968), Radford et al. (1974), and Benson (1979), as well as the botanical dictionaries such as Cook (1968), Little and Jones (1980), and Stearn (1966). The use of inflorescence terms generally follows Rickett (1944, 1955), and fruit definitions are from Judd (1985).

A

abaxial - (= dorsal). The side of an organ facing away from the axis (e.g., the underside of a leaf).

acaulescent - Stemless or apparently stemless (main stem underground).

accrescent - Increasing in size with age, after flowering.

accumbent - Describes cotyledons when the edges of both are adjacent to the radicle (*I:* Fig. 12, 5a; pg. 31).

achene - A more or less small, indehiscent, dry fruit with a thin and close-fitting wall surrounding the single seed (*I:* Fig. 39, 2; pg. 90).

actinomorphic - (= radially symmetrical, regular). Divisible into equal halves by two or more planes (*I:* Fig. 2, 1a; pg. 6).

adaxial - (= ventral). The side of an organ facing toward the axis (e.g., the upper side of a leaf).

adherent - (= connivent). Touching closely (more or less adhesively) but not fused to a dissimilar organ.

adnate - United to an unlike organ or part.

adventitious roots - Secondary roots appearing from any part of the plant (e.g., stem or leaf) other than the root system.

aestivation - The arrangement of perianth parts in the bud.

aggregate fruit - The product of several separate carpels of a single gynoecium that stick together or are connate (*I:* Fig. 3, n; pg. 10).

alternate - An arrangement of structures (such as leaves) singly at different heights along the stem, i.e., one leaf per node (*I:* Fig. 6, 1a, pg. 18).

ament - (= catkin). Any pendent, flexuous, and spike-like inflorescence of reduced, often anemophilous flowers (*I:* Fig. 20, 1i; pg. 49).

amphitropous - Describes an ovule whose funiculus (stalk) is curved around it so that the ovule tip and funiculus base are near each other (*II:* Fig. 37, 3b; pg. 92).

anatropous - Describes an ovule whose body is fully inverted and adnate to the funiculus, so that the micropyle is basal, adjoining the funiculus (*II:* Fig. 37, 3c; pg. 92).

androecium - Collective term for the "male" reproductive units (stamens) of the flower.

anemophilous - Wind-pollinated.

anther - The pollen-producing apical portion of the stamen (*II:* Fig. 37, 1b; pg. 92).

anthesis - Generally, the time when the flower expands or opens.

apical placentation - Placentation (attachment of ovules) at the distal (apical) end of the ovary.

apiculate - Terminated by a short, sharp, flexible point.

apocarpous - With the carpels separate.

apomixis - Reproduction that is apparently the result of normal sexual union but does not involve fertilization.

apopetalous - With the petals separate.

appressed (adpressed) - Closely and flatly pressed against.

arcuate - Arching or curved.

areole - Generally, a small clearly marked space; in the Cactaceae, a specialized cushion-like area on the stem that the bears a tuft of hairs, spines, and/or glochids (*II:* Fig. 9, c; pg. 22).

aril - A large and specialized outgrowth of the seed growing out of the funiculus at the hilum (attachment area of seed to funiculus).

auricle - An ear-shaped projection.

awn - A long bristle.

axil of leaf - The upper angle between a leaf and the stem.

axile placentation - Placentation (attachment of ovules) at or near the center of a compound ovary on the inner angle formed by the septa (partitions); (*I:* Fig. 13, 1j; pg. 34).

axillary - In the axil.

B

baccate - Fleshy or pulpy.

banner - (= standard). Usually the upper and largest petal of the papilionaceous flowers present in many of the Fabaceae (Faboideae); (*I:* Fig. 17, 1c; pg. 43).

basal placentation - Placentation (attachment of ovules) at the proximal (basal) end of the ovary (*I:* Fig. 10, 1e; pg. 27).

basifixed - Attached at the base.

basipetal - Developing in a longitudinal plane from an apical or distal point toward the base.

berry - An indehiscent fleshy fruit with few to many seeds (rarely a single seed); the flesh may be more or less homogeneous or heterogeneous with the outer portion more firm or leathery; septa are present in some (*I:* Fig. 26, 1h and i; pg. 64).

bicollateral bundles - A bundle having phloem on two sides of the xylem.

bifid - Two-cleft.

bilabiate - Two-lipped; often applied to the corolla or calyx (*I:* Fig. 25, 1c; pg. 62).

bilaterally symmetrical - (= zygomorphic). Divisible into equal halves in one plane only, usually along an anterior-posterior line (*I:* Fig. 2, 1d; pg. 6).

biseriate - In two whorls.

bract - A much reduced leaf often associated with flowers.

bristle - A stiff hair.

budding - A form of vegetative reproduction in which a new individual develops as an outgrowth of a mature plant.

bulb - A thickened underground bud (in a resting state) composed of a reduced stem and scales (leaf bases); (*I:* Fig. 31, 2b; pg. 74).

C

caducous - Falling off very early.

caespitose (cespitose) - Growing in tufts or mats.

calyptra - A cap, cover, or lid.

calyx - The outermost whorl of the typical perianth, composed of the sepals (*II:* Fig. 37, 1b; pg. 92).

campanulate - Bell-shaped.

campylotropous - Describes an ovule curved by uneven growth so that the micropyle is near the funiculus, but the ovule is not adnate to the funiculus (*II:* Fig. 37, 3e; pg. 92).

capitate - In a head (dense cluster of flowers).

capitulum - (= head). A compact inflorescence composed of a very short (often discoid) axis and usually sessile flowers (*I:* Fig. 29, 1b; pg. 70).

capsule - A dry fruit derived from a two- to many-carpellate gynoecium that opens (in various ways) to release few to many seeds (*I:* Fig. 7, 1e; pg. 20).

carpel - The foliar, ovule-bearing unit of a flower that forms either all (simple pistil) or part of a pistil (compound pistil).

carpellate flower - A "female flower", one having pistils and no functional stamens.

carpophore - The wiry stalk (primarily of carpellary origin) that supports each half of the dehiscing schizocarp present in some of the Araliaceae (*I:* Fig. 28, 1d; pg. 68).

caruncle - A fleshy outgrowth from the integuments situated at the attachment point (hilum) of a seed (*I:* Fig. 15, 2i; pg. 38).

caryopsis - (= grain). A more or less small, indehiscent, dry fruit with a thin wall surrounding and fused to the single seed (*I:* Fig. 41, 2i; pg. 95).

catkin - See **ament**.

caudate - Bearing a tail-like appendage.

caudicle - A portion of the pollinium of some orchids that is usually slender and composed of viscin with some pollen grains (*I:* Fig. 34, 1d; pg. 79).

cauliflorous - Describes a plant that apparently produces flowers directly upon woody branches or trunks.

cauline - Pertaining or belonging to the stem.

chaffy - Thin, dry, and membranous.

circinate - Coiled at the tip, with the apex nearest the center of the coil (*II:* Fig. 26, 2b; pg. 65).

circumscissile capsule - (= pyxis). A capsule that dehisces by a horizontal line around the fruit, the top coming off as a lid (*I:* Fig. 10, 3d; pg. 27).

clavate - Club-shaped.

clawed - A petal, sepal, or tepal that is narrowed abruptly basally into a long slender stalk (*I:* Fig. 12, 1d; pg. 31).

cleistogamous flowers - Small, closed, self-fertilized flowers (*I:* Fig. 11, 2; pg. 29).

coherent - (= connivent). Touching closely (more or less adhesively) but not fused to a similar organ.

colleter - A multicellular hair producing a sticky secretion.

column - The structure formed by the union of stamens, style, and stigma present in many of the Orchidaceae (*I:* Fig. 33, h and i; pg. 77).

coma - A tuft of hairs (*I:* Fig. 23, 1j; pg. 57).

commisure - The face by which adjoining carpels (some Araliaceae) or stigmas or style branches (Brassicaceae) are appressed.

comose - With a tuft of hairs.

compound leaf - A leaf composed of two or more leaflets (*I:* Fig. 18, 1c; pg. 45).

conduplicate - Folded together lengthwise; in particular, describes cotyledons when their dorsal sides are folded lengthwise over the radicle and more or less enclose it (*I:* Fig. 12, 1j; pg. 31).

connate - United to a similar organ or part.

connective - The tissue connecting the two locules of an anther.

connivent - See coherent.

contorted - Twisted.

contractile root - A specialized type of root, often found in bulbous plants, that contracts in length and pulls the bulb (or shoot) deeper into the soil.

convolute aestivation - Describes the perianth in bud when one edge overlaps the next part (sepal or petal or lobe) while the other edge or margin is overlapped by the preceding part.

coriaceous - Leathery.

corm - A solid bulb-like part of the stem, usually subterranean, sometimes bearing small scale leaves (*I:* Fig. 36, 1b; pg. 83).

corolla - The inner whorl (or several whorls) of the typical perianth, composed of the petals.

corona - Generally, any appendage between (or from) the corolla and/or androecium (*I:* Fig. 32, 1; pg. 75).

corpusculum - The connection ("gland") between the arms of pollinia present in many of the Apocynaceae (*I:* Fig. 23, 1g; pg. 57).

corrugated - Irregularly folded or wrinkled.

corymbose - Refers to an inflorescence that is short, broad, and flat-topped (*I:* Fig. 29, 2a; pg. 70).

costapalmate - A palmate leaf in which the petiole continues through the blade as a distinct midrib (*I:* Fig. 37, 2b; pg. 85).

cotyledon - The primary leaf in the embryo.

cruciform - Arranged diagonally (in a cross).

cupule - A cup-like structure at the base of some fruits (*I:* Fig. 19, 1j; pg. 47).

cyathium - A type of specialized inflorescence of some of the Euphorbiaceae in which the reduced unisexual flowers are congested within a bracteate envelope (*I:* Fig. 15, 1b and c; pg. 38).

cyme - Generally, a determinate, compound, and more or less flat-topped inflorescence; the basic cymose unit is a three-flowered cluster composed of a peduncle bearing a terminal flower and below it, two bracts with each bract subtending a lateral flower (*II:* Fig. 26, 1a; pg. 65).

cystolith - A crystalline concretion (usually calcium carbonate) contained within a specialized cell (lithocyst); (*I:* Fig. 14, 1d; pg. 36).

D

deciduous - Falling off at the end of each growing season.

decompound - More than once compound (or divided).

decurrent - Extending downward, as when the leaf bases extend down and are adnate to the stem, or when stigmas extend down the style (*I:* Fig. 3, m; pg. 10).

decussate - Opposite leaves alternating in pairs at right angles (*I:* Fig. 25, 1a; pg. 62).

dehiscence - The process of splitting open at maturity.

deliquescent - Describes perianth parts that quickly become semiliquid.

dentate - With angular teeth projecting at right angles to the margin of the structure (e.g., leaves).

determinate - An inflorescence in which the terminal or central flower opens first resulting in the cessation of primary axis elongation.

diadelphous - Applied to stamens when the androecium comprises two bundles (e.g., many legumes with "9+1" stamens); (*I:* Fig. 17, 3d; pg. 43).

didynamous - Stamens in two pairs, the pairs of unequal lengths (*I:* Fig. 25, 1f; pg. 62).

dioecious - Having staminate and carpellate flowers on different plants.

disk flower - The tubular flowers of many of the Asteraceae (*I:* Fig. 29, 2c; pg. 70).

distichous - Two-ranked (e.g., leaves).

distinct - Separate from parts in the same series.

dorsal - See **abaxial**; also, with thallose plants (e.g., *Lemna*), refers to the upper surface away from the substrate.

dorsifixed - Attached at the back.

drupe - An indehiscent fleshy fruit in which the outer part is more or less soft and the center has one or more hard stones (pyrenes) enclosing the seeds (*I:* Fig. 6, 1i and j; pg. 18).

E

echinulate - Covered with small bristles.

elliptical - Oval in outline, with the widest point at or about the middle (*I:* Fig. 1, 1a; pg. 4).

emarginate - With a shallow notch at the apex.

embryotega - An outgrowth on the seed coat present in many of the Commelinaceae.

enation - An epidermal outgrowth.

endosperm - The starch and oil containing tissue of many seeds derived from the triple fusion nucleus of the embryo sac.

ensiform - Sword-shaped.

entire - A margin without any toothing or division.

entomophilous - Insect-pollinated.

epicalyx - An involucre outside the true calyx (*I:* Fig. 13, 1c; pg. 34).

epigynous flower - A flower with the sepals, petals, and stamens apparently arising upon the ovary (actually growing from the edge of the floral cup, which is adnate to the ovary; ovary inferior); (*I:* Fig. 21, 4b; pg. 51).

epipetalous - Arising from the corolla.

equitant - Describes leaves that are flattened in two overlapping ranks (or rows), as in *Iris;* such leaves are often also sharply folded along their midribs (*II:* Fig. 32, 1a; pg. 80).

exstipulate - Without stipules.

extrorse - Anther dehiscence with the locule openings facing outward from the center of the flower.

F

falls - In an *Iris* flower, the outer whorl of tepals, which are spreading or deflexed and often broader than the inner tepals (*II:* Fig. 32, 2b; pg. 80).

fasciculate - Congested into bundles or clusters.

fenestration - A window-like opening or perforation.

filament - The stalk of a stamen (*II:* Fig. 37, 1b; pg. 92).

filiform - Long and slender; thread-like.

flabellate - Fan-like.

floral cup - (= hypanthium). The cup-like structure usually derived from the fusion of the perianth bases and androecium, and on which the perianth and stamens are seemingly borne (*I:* Fig. 21, 2b; pg. 51).

florets - Small individual flowers, especially those of the Poaceae (flower plus palea and lemma) and Asteraceae.

follicle - A dry to less commonly fleshy fruit derived from a single carpel that opens along a single suture (*I:* Fig. 3, n; pg. 10).

free - Separate from unlike organs or parts.

fruit - A matured ovary with associated accessory parts.

fugacious (fugaceous) - Falling or withering away very early.

funiculus - (plural: funiculi). The stalk of an ovule.

funnelform - Funnel-shaped.

G

geniculate - Bent like a knee.

gibbose - Swollen on one side.

glabrous - Without hair.

glaucous - Covered with a whitish substance that rubs off (waxy bloom).

globose - Spheroid.

glochid - A minute barbed bristle (of many cacti), often occurring in tufts (*II:* Fig. 9, d; pg. 22).

glume - A small chaffy bract; in particular, the sterile bracts at the base of most grass spikelets (*II:* Fig. 37, 2a and b; pg. 92).

grain - See **caryopsis**.

gynobasic style - A style arising directly from the receptacle and appearing to be inserted at the base of the ovary (*I:* Fig. 25, 1g and h; pg. 62).

gynoecium - A collective term for the "female" reproductive units (pistils) of the flower.

gynophore - A special stalk supporting a pistil.

gynostegium - The organ formed by the adnation of stamens and stigma present in many of the Apocynaceae (*I:* Fig. 23, 1e; pg. 57).

H

habit - The general appearance of a plant.

halophilous - Describes plants ("halophytes") that grow in soils with a high percentage of inorganic salts.

hastate - Arrowhead-shaped, with widely divergent basal lobes.

hastula - The woody ligule of many palms (*I:* Fig. 37, 1b; pg. 85).

head - See **capitulum**.

helicoid cyme - A coiled cyme in which the lateral branches develop from the same side of the main axis (*II:* Fig. 26, 1c; pg. 65).

hemitropous (hemianatropous) - Describes an ovule in which the funiculus is more curved than in an anatropous ovule (*II:* Fig. 37, 3d; pg. 92).

hesperidium - A leathery-skinned berry with several to many partitions (*I:* Fig. 16, 1f and g; pg. 41).

heterogamous head - A head with more than one type of floret.

heterostylous - Refers to plants in which the length of the mature style, relative to other parts of the flower, differs in flowers of various plants of the same species.

hilum - In a seed, the scar indicating the point of attachment of the funiculus (stalk); (*II:* Fig. 6, m; pg. 15).

hirsute - With coarse hairs rough to the touch.

hispid - With rigid or stiff hairs.

homogamous - Descriptive of a flower with the stigma receptive to pollen at the same time pollen is shed from the anthers of the same flower.

homogamous head - A head with only one kind of floret.

hood - A hood-like outgrowth of the filaments present in many of the Apocynaceae (*I:* Fig. 23, 1e; pg. 57).

horn - A projecting outgrowth of the filaments present in many of the Apocynaceae (*I:* Fig. 23, 1e; pg. 57).

husk - An outer covering of some fruits that is usually derived from the perianth and/or involucre (*II:* Fig. 17, 1l and m; pg. 43).

hyaline - Transparent, thin, and membranous.

hypanthium - See **floral cup**.

hypogynous flower - A flower with the perianth and stamens arising from below the ovary (ovary superior); (*II:* Fig. 37, 1b; pg. 92).

I

imbricate - Overlapping, like the shingles on a roof.

imperfect - A flower lacking either androecium or gynoecium.

incumbent - Describes cotyledons when their dorsal sides are parallel to the radicle (*I:* Fig. 12, 3b; pg. 31).

indeterminate - An inflorescence in which the lowermost or outermost flower opens first with the primary axis often elongating as the flowers develop; usually no terminal flower is produced.

induplicate - Folded inwards (V-shaped in cross section).

inferior - An ovary adnate to the floral

cup and thus appearing to be below the perianth (*I:* Fig. 21, 4b; pg. 51).

inflorescence - The arrangement of flowers on the floral axis or a flower cluster.

integument - The outer coating (or coatings) of an ovule that becomes the seed coat.

internode - The portion of the stem between two nodes.

introrse - Anther dehiscence with the locule openings facing inward toward the center of the flower.

involucre - A series of bracts surrounding a flower or inflorescence.

J

jaculator - (= retinaculum). A hook-like appendage on the funiculus of certain ovules (in the Acanthaceae), which aids in the expulsion of seeds from the fruit (*II:* Fig. 24, 2d and e; pg. 60).

K

keel - The two front (often lower) petals of the papilionaceous flower present in many of the Fabaceae (Faboideae); (*I:* Fig. 17, 1d; pg. 43).

L

labellum - The lip of an orchid corolla (*I:* Fig. 33, e; pg. 77).

lacerate - Torn; cut irregularly.

lamellate placentation - A modification of parietal placentation in which the ovules are attached to plate-like lamellae within the ovary.

laminar - Flat and expanded; blade-like.

lanceolate - Lance-shaped; much longer than broad and widening above the base then tapering to the apex.

latex - A colorless to more often white, yellow, or reddish sap of some plants.

laticifer - A cell or cell series containing the fluid latex.

latrorse - Anther dehiscence with the locule openings located on the sides of the anther.

legume - A usually dry fruit derived from a single carpel that opens along two longitudinal sutures (*I:* Fig. 18, 8; pg. 45).

lemma - In many of the Poaceae, the lower of the two bracts immediately

subtending the flower (*II:* Fig. 37, 2b; pg. 92).

lenticels - Spongy areas in the cork of stems (and other plant parts) that allow interchange of gases between the interior and the exterior of the stem.

lenticular - Lens-shaped; biconvex.

lepidote - Covered with small scales.

liana - A climbing woody vine.

ligulate floret - (= ray floret). A flower with a strap-shaped corolla present in many of the Asteraceae (*I:* Fig. 29, 3b; pg. 70).

ligule - A projection from the top of the leaf sheath present in many of the Poaceae and Arecaceae (*I:* Fig. 41, 2c; pg. 95).

limb - The expanded flat part of a petal or sympetalous corolla.

linear - Long and narrow, with the sides nearly parallel.

lip - One of the (usually two) parts of an unequally divided calyx or corolla (*I:* Fig. 25, 1c and d; pg. 62).

lithocyst - A cell containing a cystolith.

locule - Generally, a compartment or cavity of an organ such as an ovary or anther.

loculicidal capsule - A capsule splitting between the septa and into the locules (chambers) of the ovary (*I:* Fig. 11, 1k; pg. 29).

lodicules - Two (or three) small scale-like structures at the base of a grass flower that represent a modified perianth (*I:* Fig. 41, 2g; pg. 95).

loment - A usually dry fruit derived from a single carpel that breaks transversely into one-seeded segments (*I:* Fig. 18, 3; pg. 45).

lorate - Strap-shaped.

M

mangrove - Generally applied to several groups of tropical trees that grow in tidally flooded ground along coastal banks and are characterized by branches that spread and send down roots, thus forming multiple trunks and causing a thick growth.

marcescent - Withering, with the remains persisting.

-merous - Suffix denoting parts or numbers of each kind or series in a flower.

mesocarp - The middle layer of the pericarp (ovary wall).

micropyle - The opening between the integuments into an ovule.

monadelphous - Applies to stamens united into one group by the connation of their filaments (*I:* Fig. 13, 1g; pg. 34).

moniliform - Appearing like a string of beads.

monocolpate - A primitive type of pollen grain having a single elongate germinal furrow (*I:* Fig. 3, k; pg. 10).

monoecious - Having staminate and carpellate flowers on the same plant.

monopodial - Describes a stem in which the height is increased due to the action of a single apical meristem.

monotypic - A taxon containing only one immediate subordinate taxon, as a genus with only one species.

mucilage - A slimy and/or sticky substance.

multilacunar node - A node in a stem having numerous gaps and numerous leaf traces related to one leaf (*I:* Fig. 3, c; pg. 10).

multiple fruit - (= syncarp). A fruit formed from several flowers (and associated parts) more or less coalesced into a single structure having a common axis (*II:* Fig. 15, 1i and j; pg. 38).

multiseriate - Many-layered.

muricate - Rough due to small hard protuberances.

mycorrhiza - The symbiotic association of fungi and roots of higher plants.

N

nectary - A nectar-secreting gland.

node - The portion of the stem where a leaf arises.

nodose - Knobby.

nucellus - The inner part (megasporangium) of an ovule in which the embryo develops.

nut - A more or less large, indehiscent, dry fruit with a thick and bony wall surrounding the single seed (*I:* Fig. 19, 2b; pg. 47).

O

oblique - With the sides unequal or asymmetrical at the base.

obovate - Ovate, but with the narrower part towards the attachment (*I:* Fig. 6, 3; pg. 18).

obovoid - A three-dimensional object that is egg-shaped.

ochrea (ocrea) - A nodal sheath or tube formed by the fusion of two stipules present in members of the Polygonaceae (*I:* Fig. 9, 1b; pg. 24).

oil body - See **caruncle**.

operculate - With a lid or cap.

opposite - Two parts (e.g., leaves) occurring at the same level and on opposite sides of the axis, i.e., two leaves per node (*I:* Fig. 24, 1a; pg. 36).

orthotropous - An erect ovule with the micropyle at the apex and the hilum at the base (*II:* Fig. 37, 3a; pg. 92).

ovary - The basal portion of the pistil that bears the ovules (*II:* Fig. 37, 1b; pg. 92).

ovate - Egg-shaped in outline, with the broader end below the middle and nearer the point of attachment (*I:* Fig. 7, 2b; pg. 20).

ovule - A structure in seed plants enclosing the female gametophyte and composed of the nucellus, one or two integuments, and funiculus; differentiates into the seed after fertilization.

P

pale - A receptacular chaffy bract subtending a floret in some of the Asteraceae.

palea - In many of the Poaceae, the upper of the two bracts immediately subtending the flower (*II:* Fig. 37, 2b; pg. 92).

palmate - Descriptive of leaves and/or venation when the lobes, divisions, or veins arise at the same point (digitate); (*I:* Fig. 37, 1a; pg. 85).

paniculate - In general, a loosely and much-branched inflorescence (ultimate units may be of various types); (*I:* Fig. 41, 2a; pg. 95).

papilionaceous - The "butterfly-like" corolla type (with standard, wings, and keel) of the subfamily Faboideae (Papilionoideae) of the Fabaceae (*I:* Fig. 17, 1a; pg. 43).

papilla - (plural: papillae). A minute, rounded projection.

pappus - The specialized and reduced calyx of the Asteraceae composed of hairs, bristles, awns, or scales (*I:* Fig. 29, 2c; pg. 70).

parallel venation - Venation in which veins lie more or less parallel to the leaf margins.

parietal - The placentation type when the ovules are borne on the walls (or extrusions of the wall) within a simple or compound ovary (*I:* Fig. 11, 1i; pg. 29).

pectinate - Comb-like, as in leaves with very close and narrow divisions.

pedicel - The stalk of a flower.

peduncle - The stalk of a flower cluster.

pellucid - Clear and almost transparent in transmitted light.

peltate - Umbrella-like; attached to the stalk near the center of the lower surface.

pepo - The specialized berry of many of the Cucurbitaceae, characterized by a hard or leathery rind (epicarp) and fleshy inner tissue (with no septa).

perfect - Descriptive of a flower with both functional stamens and functional pistils.

perianth - A collective term for the calyx and corolla (or tepals).

pericarp - The wall of a fruit.

perigynium - The modified sac-like bract surrounding the achene of *Carex* (*I:* Fig. 39, 3b and c; pg. 90).

perigynous flower - A flower with the sepals, petals, and stamens arising from a floral cup that is not adnate to the ovary (*I:* Fig. 21, 3b; pg. 51).

perisperm - Seed storage tissue similar to endosperm but derived from the nucellus.

persistent - Remaining attached.

personate - Describes a bilabiate (two-lipped) corolla when the lower lip has a conspicuous rounded projection that closes the throat (*I:* Fig. 24, 3; pg. 59).

petal - One member of the inner floral envelope (corolla) of a typical flower; usually colored and more or less showy (*II:* Fig. 37, 1b; pg. 92).

petiole - The stalk of a leaf.

petiolule - The stalk of a leaflet.

phyllary - An involucral bract present in many of the Asteraceae.

pinna - (plural: pinnae). A primary division or leaflet of a pinnately compound leaf.

pinnate - Refers to a leaf in which there are more than three leaflets arranged in two rows along a common axis.

pinnate venation - Venation consisting of a central midvien with many secondary veins emerging on both sides to form a feather-like pattern.

pistil - The ovule-bearing organ of the flower composed typically of at least one ovary, a style (when present), and a stigma (thus formed from one or more carpels); (*II:* Fig. 37, 1b; pg. 92).

placenta - (plural: placentae). The place or part in the ovary where the ovules are attached.

placentation - The arrangement of ovules within the ovary.

plicate - Folded into plaits, as in a fan.

plumose - Describes feather-like compound hairs.

plynerved - A leaf with two major secondary veins arching toward the apex.

pollen grains - The microgametophytes, which contain the male generative cells, borne in the anthers of the flower.

pollinium - The coherent waxy pollen mass produced by many members of the Orchidaceae and Apocynaceae; transported as a unit during pollination (*I:* Fig. 33, m; pg. 77).

polycolporate - A type of pollen grain having many (compound) apertures.

polygamodioecious - Describes a species which is functionally dioecious but has a few flowers of the opposite sex or a few perfect flowers at flowering time.

polygamous - Bearing perfect and imperfect flowers on the same plant.

polyphyletic - Applied to a composite taxon derived from two or more ancestral sources; not of a single immediate line of descent.

pome - An indehiscent fleshy fruit in which the outer part is more or less soft and the center has papery or cartilaginous structures enclosing the seeds (*I:* Fig. 21, 4d and e; pg. 51).

precocious - Describes flowers that appear early in the season (before the leaves).

prickle - A sharp, pointed outgrowth from the epidermis or cortex.

prophyll - A small bract.

protandrous (proterandrous) - Descriptive of a flower with the anthers shedding the pollen before the stigma of the same flower is receptive.

protogynous (proterogynous) - Descriptive of a flower with the stigma receptive to pollen before pollen is shed from the anthers of the same flower.

pseudanthium - A cluster of small or reduced flowers that collectively simulate a single flower (*I:* Fig. 29, 1a; pg. 70).

pseudobulb - The thickened bulb-like stem of certain orchids (*I:* Fig. 33, a; pg. 77).

pubescent - Covered with hairs.

pulvinus - An enlargement at the petiole base at its point of attachment to the stem or at the petiole apex at its point of attachment to the blade (*I:* Fig. 18, 1b; pg. 45).

punctate - Dotted with translucent or colored glands, dots, or depressions (*I:* Fig. 7, 2b; pg. 20).; also describes a structure that is small and roughly circular.

pyrene - A stone of a drupe composed of a seed and a bony endocarp (inner wall of the fruit); (*I:* Fig. 6, 1k; pg. 18).

pyxis - See **circumscissile capsule**.

R

racemose - Applies to an inflorescence having stalked flowers (or small flower clusters) arranged along an elongated central axis (*I:* Fig. 33, a; pg. 77).

rachilla - In many of the Poaceae and Cyperaceae, the axis that bears the florets (*II:* Fig. 37, 2b; pg. 92).

radial symmetry - See **actinomorphic**.

radicle - The embryonic root of an embryo or a germinating seed.

raphe - The portion of the funiculus adnate to the integument (often appearing as a seam on the seed coat).

raphide - A needle-like crystal of calcium oxalate.

ray floret - See **ligulate floret**.

receptacle - The more or less enlarged or elongated apex of the flower axis which bears some or all of the flower parts (*I:* Fig. 3, g; pg. 10).

recurved - Bent or curved downward or backward.

reduplicate - Folded downward (A-shaped in cross section).

reniform - Kidney-shaped.

replum - The persistent partition of the ovary (and fruit) present in many of the Brassicaceae (*I:* Fig. 12, 2e; pg. 31).

resupinate - Twisted 180°; turned upside down.

retinaculum - (plural: retinacula). See **jaculator**.

rhizome - An underground stem or the climbing stem of an epiphyte.

rosette - A circular cluster of leaves, usually close to the ground.

rostellum - The beak- or strap-like portion of the stigma of an orchid, specialized to take part in the transfer of pollen.

rotate - A sympetalous corolla with a flat, circular, and spreading limb at right angles to a very short tube (*I:* Fig. 24, 4; pg. 59).

ruminate - Mottled in appearance with dark and light zones (*I:* Fig. 15, 2i; pg. 38); also describes endosperm which is irregularly grooved and ridged (*I:* Fig. 4, 1j; pg. 13).

S

sagittate - Arrowhead-shaped; triangular with the basal lobes pointing downward (*I:* Fig. 35, 2; pg. 81).

salverform - Descriptive of a sympetalous corolla with a slender tube and an abruptly expanded flat limb (*I:* Fig. 23, 2a; pg. 57).

samara - A winged, indehiscent, more or less dry fruit containing a single seed (*I:* Fig. 20, 3b; pg. 49).

sarcotesta - A fleshy seed coat.

scabrous - Rough to the touch.

scalariform perforation plate - In vessel members of the xylem, a type of multi-perforate plate in which elongated perforations are arranged parallel to one another so that the cell wall bars between them form a ladder-like pattern.

scale - A general term for small, often dry, appressed leaves, bracts, or hairs.

scapose - A plant bearing flowers or inflorescences on a leafless peduncle arising from the ground.

scarious - Dry, thin, membranous, non-green, and transluscent.

schizocarp - A fruit derived from a two- to many-carpellate gynoecium that splits into two or more one-seeded segments (*I:* Fig. 15, 2g and h; pg. 38).

scorpioid cyme - A coiled cyme in which the apparently lateral flowers (or branches) develop alternately on opposite sides of the main axis (*II:* Fig. 26, 1b; pg. 65).

scurfy - With flake-like particles (small scales) on the surface.

seed - The product of the ovule after fertilization, comprising the embryo with its surrounding food reserves and protective coverings.

sensu lato - Latin, "in the broad sense", with a wide or general interpretation; abbreviated *s.l.*

sensu stricto - Latin, in a narrow sense", with a restricted interpretation; abbreviated *s.s.*

sepal - One of the outer floral envelopes (calyx) of a typical flower; usually green and foliaceous (*II:* Fig. 37, 1b; pg. 92).

septicidal capsule - A capsule splitting along the septa (partitions between the locules); (*I:* Fig. 7, 1e; pg. 20).

septifragal capsule - A capsule splitting longitudinally so that the valves break away from the septa (partition walls); (*II:* Fig. 27, m; pg. 67).

septum - (plural: septa). A partition or cross wall.

serrate - Describes a saw-toothed margin with the teeth pointing forward (*I:* Fig. 20, 2a; pg. 49).

sessile - Without a stalk.

shield cells - The cells of the expanded apical portion of the peltate hairs present on the leaves of many of the Bromeliaceae (*II:* Fig. 35, 2b; pg. 87).

silicle - A dry fruit of many of the Brassicaceae derived from a two-carpellate gynoecium in which the two valves split away from a persistent partition (around the rim of which the seeds are attached); usually not more than twice as long as wide (*I:* Fig. 12, 3a; pg. 31).

silique - Same as **silicle**, but usually at least twice as long as wide (*I:* Fig. 12, 2d; pg. 31).

simple leaf - A leaf not divided into leaflets.

spadix - A spike with a thickened fleshy axis (*I:* Fig. 36, 2a; pg. 83).

spathe - The large bract surrounding or subtending an inflorescence (*I:* Fig. 36, 1c and d; pg. 83).

spatulate - Spoon-shaped; oblong or somewhat rounded with the basal end long and tapered.

spicate - Refers to an inflorescence having sessile flowers arranged singly or in contracted clusters along a central axis (*I:* Fig. 36, 2a; pg. 83).

spike - See **spicate**.

spikelet - The small bracteate spike of many of the Poaceae and Cyperaceae (*II:* Fig. 37, 2a and b; pg. 92).

spur - A sac-like projection from a flower part, usually a petal or sepal (*I:* Fig. 34, 1b; pg. 79).

stamen - The pollen-bearing organ of the flower, composed of a filament (stalk) and anther (pollen sacs); (*II:* Fig. 37, 1b; pg. 92).

staminate flower - A "male" flower, one having stamens and no functional gynoecium.

staminode - A sterile stamen.

standard - See **banner**; also, in an *Iris* flower, one of the inner whorl of tepals that are erect or ascending and often are narrower than the outer tepals (*II:* Fig 32, 2c; pg. 80).

stellate - Star-shaped; with radiating branches.

stigma - The apical part of the pistil that is receptive to pollen (*II:* Fig. 37, 1b; pg. 92).

stipe - A strap of tissue derived from the column that connects the orchid pollinia to the viscidium.

stipule - One of a pair of appendages at the base of the petiole at the point of attachment to the stem (*I:* Fig. 22, 3b; pg. 54).

stolon - A runner; a horizontal stem at or below the ground surface that produces a new plant at its tip.

striate - With fine longitudinal lines, channels, or ridges.

stylar canal - In a few angiosperms, a canal (or one of several canals) lined with specialized tissue through the center of the style.

style - The more or less elongated portion of the pistil between the ovary and stigma (*II:* Fig. 37, 1b; pg. 92).

stylopodium - The nectariferous enlargement at the style bases present in many of the Araliaceae (*I:* Fig. 27, 1d; pg. 66).

succulent - Fleshy, soft, and thick.

superior - An ovary above the point of attachment of the perianth and androecium and free from them (*I:* Fig. 31, 1c; pg. 74).

suture - A line of joining; a seam.

syconium - In *Ficus*, the enlarged, hollow, and flask-like structure that bears the flowers and fruits along the inner surface (*II:* Fig. 15; 2a and b, 3a and b; pg. 38).

sympetalous - A corolla with united petals, at least at the base.

sympodial - Describes a stem in which the apex aborts or becomes reproductive, with the vegetative growth being continued in each instance by a lateral branch.

syncarp - See **multiple fruit**.

syncarpous - With the carpels united.

syngenesious - Describes stamens connate by their anthers to form a cylinder (*I:* Fig. 29, 2j; pg. 70).

synsepalous - A calyx with united sepals, at least at the base.

T

taproot - The primary root when markedly larger than the others.

tendril - An elongated twining segment of a leaf, stem, or inflorescence by which a plant attaches to its support.

tenuinucellate - With the nucellus consisting of a single layer of cells.

tepal - A segment of those perianths not clearly differentiated into a typical calyx and corolla.

terete - Approximately circular in cross section.

terminal - At the tip or apical end.

testa - The seed coat (hardened mature integuments).

tetrad - (adjective: tetradenous). A group of four.

tetradynamous - Describes an androecium of six stamens with the four inner longer than the outer two (*I:* Fig. 12, 1f; pg. 31).

thallus - (plural: thalli). A plant body, often flat and leaf-like, that is not clearly differentiated into roots, stems, and leaves.

thorn - A pointed reduced branch (*I:*

Fig. 21, 4a; pg. 51).

toothed - With small projections and indentations alternating along the margin.

translator arm - One of the elongated structures connecting the pollinia of adjacent anthers present in many of the Apocynaceae (*I:* Fig. 23, 1g; pg. 57).

trichome - A hair or bristle.

trifid - Three-cleft.

trifoliolate - With three leaflets.

trigonous - Three-angled.

trilacunar node - A node in a stem having three leaf gaps related to one leaf (*I:* Fig. 20, 2b; pg. 49).

truncate - Describes the base or apex of a part that ends abruptly and appears as though squarely cut off.

tuber - A thickened short underground part of the stem functioning as a storage area for reserve food (*I:* Fig. 23, 1a; pg. 57).

tubercle - The thickened persistent style base on the achenes of some of the Cyperaceae (*I:* Fig. 39, 2; pg. 90).; also a rounded protrusion (bump) on a surface.

turbinate - Top-shaped; more or less resembling an inverted cone.

U

umbel - An inflorescence composed of several branches that radiate from almost the same point and that are terminated by single flowers or secondary umbels (*I:* Fig. 27, 1a and 3; pg. 66).

unilacunar node - A node in a stem having one leaf gap related to one leaf (*I:* Fig. 5, c; pg. 16).

uniseriate - In one series or whorl.

unitegmic ovule - An ovule with a single integument.

urceolate - Urn-shaped; more or less globular with a contracted mouth and small flaring lobes (*I:* Fig. 8, 1c; pg. 22).

utricle - A more or less small, indehiscent, dry fruit with a thin wall (bladder-like) that is loose and free from the single seed (*I:* Fig. 10, 2; pg. 27).

V

valvate aestivation - When the perianth parts in bud meet by the edges without overlapping.

velamen - The absorbent and protective root epidermis of epiphytic orchids.

ventral - See **adaxial**; also, with thallose plants (e.g., *Lemna*) refers to the lower surface next to the substrate.

versatile - Descriptive of an anther attached to its filament at the middle and usually moving freely.

verticil - A whorl.

vestiture - The covering on a surface; e.g., hairs or scales.

viscidium - A sticky part of the rostellum of some orchids that serves to attach the pollinia to the pollinator (*I:* Fig. 34, 1e; pg. 79).

viscin - An elastic and somewhat viscid material.

vitta - Aromatic oil or resin canal in the pericarp of many of the Araliaceae.

viviparous - Describes seeds that germinate while still attached to the parent plant; also describes a plant that produces these seeds.

W

whorl - A group of leaves or other structures arranged in a circle at a single node (e.g., three or more leaves per node); (*I:* Fig. 25, 1a; pg. 62).

wings - Thin, often membranous extensions of a structure (*I:* Fig. 20, 3b; pg. 49); also the lateral pair of petals of a papilionaceous corolla present in many of the Fabaceae (Faboideae); (*I:* Fig. 17, 1b; pg. 43).

Z

zygomorphic - See **bilaterally symmetrical**.

References Cited

Benson, L. 1979. Plant classification. 901 pp. D.C. Heath and Co., Lexington, MA.

Cook, J.G. 1968. ABC of plant terms. 293 pp. Merrow Publishing Co., Ltd., Watford, England.

Cronquist, A. 1968. The evolution and classification of flowering plants. 396 pp. H. Mifflin, Boston, MA.

Judd, W.S. 1985. A revised traditional/descriptive classification of fruits for use in floristics and teaching. Phytologia 58: 233-242.

Lawrence, G.H.M. 1951. Taxonomy of vascular plants. 823 pp. Macmillan Publishing Co., Inc., New York, NY.

Little, R.J. and C.E. Jones. 1980. A dictionary of botany. 400 pp. Van Nostrand Reinhold Co., New York, NY.

Radford, A.E., W.C. Dickison, J.R. Massey, and C.R. Bell. 1974. Vascular plant systematics. 891 pp. Harper & Row, Publishers, New York, NY.

Rickett, H.W. 1944. The classification of inflorescences. Bot. Rev. 10: 187-231.

_____. 1955. Materials for a dictionary of botanical terms. III. Inflorescences. Bull. Torrey Bot. Club 82: 419-445.

Stearn, W.T. 1966. Botanical Latin. 566 pp. Hafner Publishing Co, New York, NY.

INDEX TO FAMILIES AND ILLUSTRATED GENERA

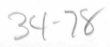